Constrained Supply and Production Planning in SAP APO

Shaun Snapp

Constrained Supply and Production Planning in SAP APO

For information about this title or to order other books and/or electronic media, contact the publisher:
SCM Focus Press
PO Box 29502 #9059
Las Vegas, NV 89126-9502
http://www.scmfocus.com/scmfocuspress
(408) 657-0249

ISBN: 978-1-939731-06-7

Printed in the United States of America

Cover and Interior design by: 1106 Design

Contents

CHAPTER 1

Introduction

There are books that explain constrained supply planning, and books that explain constrained production planning. However, I am unaware of a book that really focuses, at a detailed level, on these topics and on how constrained supply and production planning work together. While there are a number of consultants who can tell you how constraints will work in very specific ways with supply planning or within the production planning application, understanding—at the deepest level—how the constraints integrate across the two applications is rare. Obtaining a detailed understanding is one of the focuses of this book. I can say that the story sounds better at a higher level than at a more in-depth view. In fact, gaps appear once you take the analysis down to the most detailed level. It took me many years to gain this understanding, but by reading this book, anyone can obtain this knowledge much more quickly and easily than it came to me.

What Is and Is Not Covered in this Book

I don't cover all of the techniques that SAP says are standard functionality in SNP and PP/DS. Much of the standard functionality with respect to constraint-based planning either does not work or is so high maintenance that it is not used in the field. With respect to resources specifically, and capacity planning in general, SAP has quite a few areas of functionality that are really just fluff and don't have any practical use. Two good examples of this are repetitive manufacturing functionality and bottleneck resource functionality. I discuss both of these overrated areas of functionality in this book.

To those hoping that the next version of SAP will fix all the problems, I say, "Don't be so optimistic." It must be understood that SAP sales dominate SAP consulting, so "implement-ability" is not a priority; sales is the priority. Individuals who write from the reverse perspective (i.e., that consulting is a higher priority than sales) produce more accurate material. Companies that take SAP's word (or the word of large consulting companies, which are sales arms for SAP) for what will work in the future will end up with solutions that function poorly and an IT budget that is wasted in the process. Many of these functionality areas are relatively useless and exist primarily so that sales can say "yes we have this functionality" and obtain the sale. In most cases this functionality that I am describing has been inoperative for years and is likely to stay that way. Those that say "SAP has to fix this," do not understand the monopolizing nature of the enterprise software market or SAP's dominant position in it. The following sub-site is dedicated to explaining the enterprise software market:

http://www.scmfocus.com/enterprisesoftwarepolicy/

I have been working in APO since 2003 and in SAP since 1997, and have seen many projects during that time. As such, I feel confident in these views on this topic, as the required data points support them and I have performed the necessary triangulation with other consultants. Almost every author in SAP either works for SAP, is unable to criticize SAP because they work for an organization that is associated with SAP, is simply not interested in dealing with controversy, or would be censored by any publishing company that they work with. Therefore, almost every author I have read (and I have read all the books on SAP APO, and many others in SAP generally) write from the perspective of what SAP *can do in theory*—that is, writing almost from the sales perspective rather than writing what SAP actually does in practice, or writing from the implementation perspective.

This book describes what *is* rather than what SAP would like to be, as has always been an objective of SCM Focus. We are fairly dominant in the domain of reality-based technical writing on SAP, although it's not hard to dominate when no one else is interested in competing. As an independent consultant with no indebtedness to SAP, I can and I do write from a realistic implementation perspective in this book. This independence gives me all kinds of flexibility that most other authors do not enjoy.

Companies could save enormous amounts of money, if, rather than relying on SAP's and the consultants' misrepresentation of what SAP can do, they knew how SAP is implemented. However, this information is very hard to come by, so new clients repeat the same bad decisions, even though the story on the functionality has been clear to me for years.

The Use of Screen Shots in the Book

I consult in some popular and well-known applications, and I've found that companies have often been given the wrong impression of an application's capabilities. As part of my consulting work, I am required to present the results of testing and research about various applications. The research may show that a well-known application is not able to perform some functionality well enough to be used by a company, and point to a lesser-known application where this functionality is easily performed. Because I am routinely in this situation, I am asked to provide evidence of the testing results within applications, and screen shots provide this necessary evidence.

Furthermore, some time ago it became a habit for me to include extensive screen shots in most of my project documentation. A screen shot does not, of course, guarantee that a particular functionality works, but it is the best that can be done in a document format. Everything in this book exists in one application or another, and nothing described in this book is hypothetical.

Disclaimer on the Field Definitions

This book in parts has a large number of field definitions. As with my book *Planning Horizons, Calendars and Timings in SAP APO* (which had an enormous number of field definitions), I needed some portions of the SAP Help definitions, but I did not want to use all of SAP's full or complete definitions, as I wanted to be able to weave in my viewpoint on the definition without making the reader continually stop and start between SAP's definition and my definition. Therefore, I settled on using the portions of the field definitions from SAP, but in combination with my shorthand. Some of the definitions are quite lengthy, so I have tried to provide a synopsis of them and add some translation where necessary. I did not want to be concerned with separating out the exact SAP quotations from my text, so I am putting this disclaimer at the beginning of this book. ***I am not trying to take credit for SAP's work, or say that any of the field definitions in this book are original.*** I could have included the full definition of each field in this book, but that would have made the book quite tedious. That is why I think providing a synopsis is more valuable to readers. The definitions are the starting point; I have a layer of analysis and learning aids such as graphics that help clarify many of the time settings. Finally, I have not included every timing field in every category. I have covered most of them, but some are very infrequently used.

Timing Field Definitions Identification

This book is filled with lists. Some of these lists are field definitions. To help you quickly identify field definitions, all text in the field definition is *italicized*. In lists that are not field definitions, only the term defined is *italicized*, while the definition that follows is not italicized.

How Writing Bias Is Controlled at SCM Focus and SCM Focus Press

Bias is a serious problem in the enterprise software field. Large vendors receive uncritical coverage of their products, and large consulting companies recommend the large vendors that have the resources to hire and pay consultants rather than the vendors with the best software for the client's needs.

Just as in my consulting practice, I do not financially benefit from a company's decision to buy an application that I showcase in print, either in a book or on the website. SCM Focus has the most stringent rules related to controlling bias and restricting commercial influence of any information provider. These "writing rules" are provided in the link below:

http://www.scmfocus.com/writing-rules/

If other information providers followed these rules, I would be able to learn about software without being required to perform my own research and testing for every topic.

Information about enterprise supply chain planning software can be found on the Internet, but this information is primarily promotional or written at such a high level that none of the important details or limitations of the application are exposed; this is true of books as well. When only one enterprise software application is covered in a book, one will find that the application works perfectly; the application operates as expected and there are no problems during the implementation to bring the application live. This is all quite amazing and quite different from my experience of implementing enterprise software. However, it is very difficult to make a living by providing objective information about enterprise supply chain software, especially as it means being critical at some point. I once remarked to a friend that SCM Focus had very little competition in providing untarnished information on this software category, and he said, "Of course, there is no money in it."

The Approach to the Book

By writing this book, I wanted to help people get exactly the information they need without having to read a lengthy volume. The approach to the book is essentially the same as my previous books, and in writing this book I followed the same principles.

1. **Be direct and concise.** There is very little theory in this book and the math that I cover is simple. While the mathematics behind the optimization methods for supply and production planning is involved, there are plenty of books that cover this topic. This book is focused on software, and for most users and implementers of the software, the most important thing to understand is conceptually what the software is doing.

2. **Use project experience.** Nothing in the book is hypothetical; I have worked with it or tested it on an actual project. My project experience has led to my understanding a number of things that are not covered in typical supply planning books. In this book, I pass on this understanding to you.

3. **Saturate the book with graphics.** Roughly two-thirds of a human's sensory input is visual, and books that do not use graphics—especially educational and training books such as this one—can fall short of their purpose. Graphics have also been used consistently and extensively on the SCM Focus website.

Before writing this book, I spent some time reviewing what has already been published on the subject. This book is different from other books in terms of its intended audience and its scope. It is directed toward people that have either worked with ERP or know what it is; I am assuming that the reader has a basic knowledge level in this area.

The SCM Focus Site

As I am also the author of the SCM Focus site, http://www.scmfocus.com, the site and the book share a number of concepts and graphics. Furthermore, this book contains many links to articles on the site, which provide more detail on specific subjects. This book provides an explanation of how supply and production planning software works and aims to continue to be a reference after its initial reading. However, if your interest in supply planning software continues to grow, the SCM Focus site is a good resource to which articles are continually added.

The SCM site dedicated specifically to supply planning is http://www.scmfocus
.com/supplyplanning.

The SCM site dedicated specifically to production planning is http://www.scm
focus.com/productionplanningandscheduling/.

The site dedicated to SAP planning is http://www.scmfocus.com/sapplanning.

Intended Audience

The feedback I received from early reviewers described this book as good for any
person who wants to understand how constraint-based planning works in supply planning and production planning in an integrated fashion as well as how
it specifically works in SAP APO. If you have any questions or comments on the
book, please e-mail me at shaunsnapp@scmfocus.com.

Abbreviations

A listing of all abbreviations used throughout the book is provided at the end of
the book.

CHAPTER 2

Understanding the Basics of Constraints in Supply and Production Planning Software

Constraint-based planning works by setting up limiting factors in a model. One of the most important limiting factors is called a resource. Resources can be anything from factory equipment to transportation units to handling equipment. These resources are then assigned limitations in capacity and/or availability (e.g., they have a specific capacity and can only be run from 8 a.m. to 10 p.m.). The goal is to match the resources as closely as possible to the realities of the environment being modeled. Unlike unconstrained (or infinite) planning, which will allocate requirements to a resource per the requests from the order management system (determined strictly by order dates), constraint-based planning moves the activity to a different period (and sometimes a different resource) when resources become filled. Assuming the resources represent reality, a plan generated in this way is feasible and implementable.

Let's define a few terms that will come up repeatedly on the topic of constraint-based planning.

1. *Constraint:* This is what restricts the capacity of the system. Most often it is a resource.

2. *Feasible:* This means the output from the planning run has met all of the constraints. The first objective is to produce a "feasible plan." For instance, when SNP creates an initial production plan, it releases a feasible plan to PP/DS.

3. *Optimal:* This is one of the most misunderstood terms in constraint-based planning, and generally is also a very overused term. "Optimal" is the second objective of constraint-based planning, but in actual fact is rarely attained. Optimal has a very specific technical meaning, which does not translate very well to the layman. Optimality is simply the condition of meeting the objective function. The objective function is the goal of any optimizer. An optimizer is expressed as the combination of an objective function along with a series of constraints. However, the actual relationship between the optimizer meeting its objective function and meeting the business requirements is usually greatly overestimated. The statement "the optimizer reached an optimal solution" does not do any of the following:

   a. It does not say whether the correct or best objective function was used. After a number of years analyzing many solutions, I have concluded that most applications are not using the best objective function. See the article at http://www.scmfocus.com/supplyplanning/2011/07/10/customizing-the-optimization-per-supply-chain-domain/.

   b. The statement does not say whether the constraints were set in a way that reflects reality, and if the company has chosen the correct constraints. This is a constant problem on projects where the technical team is quite proud of their creation, but the business outcomes of the optimizer end up being quite poor.

   c. For all the talk of optimal solutions, it is in fact rare for the SNP or PP/DS cost optimizer to solve the overall problem optimally; at most it will typically solve a percentage of the sub-problems optimally. Furthermore, the percent optimally solved does not correlate with the quality of the business solution. I have tested this extensively and correlated the percent optimally solved with key business metrics and never found a relationship.

I can give many more examples of why one should be careful when using the term "optimal" or "optimality," but I would like to move back into the core topic of this chapter.

The following image shows how constraint-based planning is expressed graphically. Only the simplest examples of optimization can be expressed in this way; once the number of constraints increases, a two-dimensional representation no longer suffices.

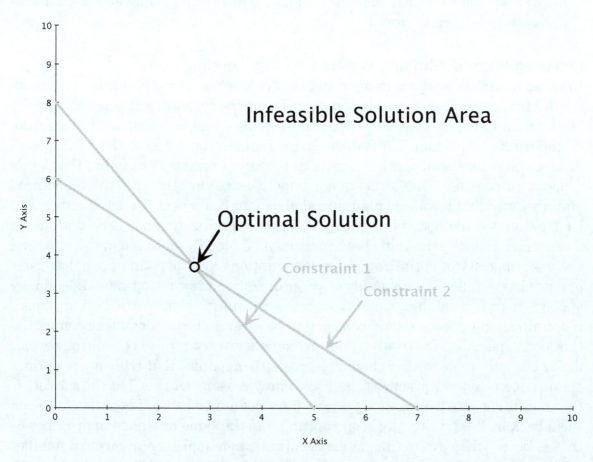

When used in planning, constraints limit the solution to what is feasible, or what is actionable, by the business. By limiting the solution, the time and processing spent evaluating recommendations that cannot be converted in reality are also limited.

Optimization is one of the constraint-based supply planning methods available in SNP and PP/DS. The other is an allocation method called Capable to Match (CTM). As with the SNP optimizer, CTM can plan while meeting all constraints, but it processes the supply or production planning problem order by order (CTM can be used for purely production planning as well as supply planning). Therefore its problem decomposition—or how it breaks the problem into smaller pieces—is entirely different from the SNP optimizer.

http://www.scmfocus.com/sapplanning/2011/10/12/snp-optimizer-sub-problem-division-and-decomposition/

Constraint-based Planning Versus Capacity Leveling

In order to understand the many benefits of constraint-based planning, it is quite helpful to compare constraint-based planning to the competing approach: capacity leveling. Capacity leveling is the much older method and was first used in an automated (rather than manual) fashion and was the second step to the MRP procedure. Constraint-based planning methods perform a one step procedure that levels capacity in one step. One of the major impetuses behind the growth of advanced planning was that advanced planning applications could perform optimization and ERP systems could not. While it may seem that optimization has been with us for a long time, in fact, it has only been implemented broadly since the mid-1990s, and the vast majority of optimization implementations are operating far below their potential. Judging from my analysis of optimization projects, optimization simply appears to be beyond the capabilities of most consulting companies, I often wonder if optimization projects should continue to be given to large consulting companies that don't specialize in the area and instead offer every type of consulting service under the sun. Instead they should be given to companies that truly make optimization a focus of their practice. There are many reasons for this, but the sad fact is that approaches to optimization projects have not evolved since I began working on them back in 1998. I literally keep running into the same mistaken approaches as those I noted when evaluating previous optimization implementations. An Achilles heel common to these types of projects is listed in the article below:

http://www.scmfocus.com/inventoryoptimizationmultiechelon/2011/05/socializing-supply-chain-optimization/

Capacity leveling uses unconstrained resources, which allow any load to be placed on any resource. Without something to constrain how demand flows through the supply and production planning during the initial supply plan (the MPS or the S&OP planning runs[1]), the demand is usually applied to resources based upon the demand data—subtracted from the supply and production lead times. Periods of high demand will lead to overloads on resources, again subtracted from the supply and production lead-time. Capacity leveling is the second step where these loads are moved to periods where there is capacity—hence the term "capacity leveling."

In contrast, constraint-based planning is a one-step planning process, while capacity leveling is a two-step process. When constraint-based planning is used, a constraint stops the overloading of resources before it occurs. The constraint is part of the initial supply and production run. There is no "second step" required, because in one step, the system accepts demands and moves the demands to periods where they can be feasibly supplied. A constraint places a hard stop on how much load (i.e., demand) can be placed upon a resource. Constraint-based planning automatically levels the load by pushing the demand forward or backward depending upon the rules that are set up for forward and backward scheduling (which is covered further on in this chapter).

Capacity leveling is used on "bottleneck resources" a topic, which I cover in Chapter 5: "Resources." I quote the following from SAP Help:

> *SNP capacity leveling is used to level specific bottleneck resources. It is run locally on a resource in a specified horizon, which means that dependencies with other resources are ignored. Leveling capacity for the entire supply chain would, in effect, be the same as performing a new planning run and is not within the scope of this function.*

Some reading between the lines is required when evaluating this quote, which states that SNP lacks the ability to group capacity level a series of resources that are in a chain. With SAP, SNP, or PP/DS capacity leveling, one can capacity-level a ***single, unconnected resource***. This answers a question that I have been

---

[1] For the purposes of simplification, I will restrict the conversation to these runs for now, and leave out deployment.

asked as to whether SNP can capacity-level across two resources that are linked in a chain, such as a liquid processing resource that then connects to a bottling resource. Leveling capacity across two resources is explained in detail in Chapter 3: "Integrating Supply and Production Software with Constraints."

Capacity leveling can be performed manually or with a capacity leveling heuristic.

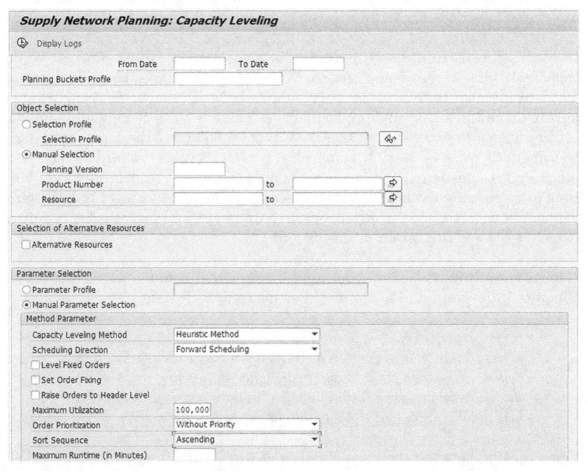

Running the SNP capacity leveling heuristic requires the user to configure a number of settings, including the following:

1. *The Capacity Leveling Method:* The options are heuristic-based capacity planning and optimization-based capacity leveling. In addition, SAP has a hook or BAdi for any custom code that the client would like to add. The

most commonly-used method for capacity leveling is by far the capacity leveling heuristic, the steps of which are listed below:

a. Step 1: Determine the Capacity Load—Moving Forward or Backward Along the Planning Horizon: *"Heuristic-based capacity leveling starts from the start or end of the planning horizon depending on the scheduling direction you chose (forward or backward), and compares the resource capacity load in each period with the required load that you defined."*

b. Step 2: Select All Overload Related Activities/Orders: *"If the system detects a resource overload, it first selects all the activities or orders that are causing the overload in the period concerned."*

c. Step 3: Sort the Overload Related Orders: *"The system then sorts these orders according to the priority you specified."*

d. Step 4: Move the Sorted Orders Forward or Backward: *"...in turn, moves orders or partial order quantities into later or earlier periods until the maximum resource capacity level has been achieved."*[2]

So as should be quite clear, prioritization is critical to the sort order and therefore to capacity leveling. You can also control the sort sequence of the orders as a following field demonstrates. However, if no priority is set, capacity leveling will simply move the overloaded orders forward or backwards without a sortation. The priority used for capacity leveling is the product priority. This priority, along with the other priorities in APO, is discussed in several of the following paragraphs.

2. *The Scheduling Direction:* Should the capacity leveling look only forward, only backwards, or both backwards and forwards to move capacity? The options are as follows:
   a. *Forward*
   b. *Backward*
   c. *Combined Forward and Backward*

3. *Level Fixed Orders?* This also levels orders that are fixed (in previous planning runs). Without this setting, the capacity leveling run is a net

---

[2] *"During forward scheduling, the system moves the orders into the future so that the first activity that uses the resource to be leveled starts after the period with the overload. During backward scheduling, the system moves the orders into the past so that the final activity that uses the resource to be leveled is completed before the start of the period with the overload."* — **SAP Help**

change operation. With this setting any planned order can be moved. The decision on this setting is typical for most settings of this type, such as the fixed or dynamic pegging setting in CTM (http://www.scmfocus. com/sapplanning/2009/05/06/pegging-in-scm/). That is, it is essentially a question of having a responsive plan versus a stable plan. One should also understand how fixing works with firming (http://www.scmfocus.com/ sapplanning/2012/06/22/firming-in-apo/).

4. *Set Order Fixing:* This fixes orders so that ordinarily they cannot be changed by a subsequent planning run. (See the previous setting for how this can be overwritten.)

5. *Raise Orders to Header Level:* Brings dependent demand resources up to the top level before running capacity leveling. This addresses one of the main limitations of traditional capacity leveling: it only levels the finished goods resources and leaves the semi-finished and component resources that are part of the finished goods bill of material (BOM) unleveled. This topic is addressed in the following section—Capacity Leveling and Dependent Demand.

6. *Maximum Utilization:* Here one can define a maximum, which is different than in the resource. So if 110% is set, then the resources that are capacity-leveled can be increased by 10% over what is set in the resource itself.

7. *Order Prioritization:* The options are as follows:
   a. Without Priority
   b. Based upon Order Size
   c. Product Priority. (This option allows capacity leveling to be performed based upon the product prioritization set in the Product Master.)

8. *Sort Sequence:* Determines if the priorities set in the Order Prioritization field (see above) are sorted in ascending or descending order. "Ascending" means that the priority sequence is 1, 2, 3... for products assigned a product priority. When the Order Size option is selected, "ascending" means the smallest orders are processed first. "Descending" means the converse for both.

9. *Maximum Runtime:* How long would you like capacity leveling to run? When the SNP leveling heuristic is used, it is quite fast. However, this setting is necessary when the optimizer is used for capacity leveling.

Capacity leveling only works in SNP for production resources and transportation resources, ***and does not work for storage or handling resources***. This is an extremely important point, and easy to forget. As a result, these resource types must be either leveled manually, or must be set as finite resources, and then either of the two constraint-based methods in SNP can be used—either CTM, or the cost optimizer to plan them. Furthermore, capacity leveling only works on four of the resource types: bucket, single-mixed, multi-mixed, and transportation. (Mixed resources are those that can be shared between SNP and PP/DS. Don't worry about understanding this topic completely yet, as this is extensively covered in Chapter 5: "Resources.") The following quote from SAP Help highlights something that is quite interesting on this topic.

> *Capacity leveling only takes into account SNP planned orders and SNP stock transfers. Deployment stock transfers, TLB shipments, and Production Planning and Detailed Planning orders (PP/DS orders) are not leveled; however, the system does take into account the resource load caused by these orders.*

What this means is that SNP capacity leveling for planned orders should only be performed outside the PP/DS planning horizon. This is because PP/DS will have created planned orders within its horizon. If SNP capacity leveling is run in a period when there are both PP/DS planned orders and SNP planned orders, SNP capacity leveling will not level the PP/DS planned orders, leading to an incomplete leveling. This is one of the issues to be worked out during the phase of the project where the time horizons and responsibilities are divided between SNP and PP/DS for production. This topic is covered in detail in Chapter 3: "Integrating Supply and Production Software with Constraints."

Capacity Leveling and Dependent Demand

Dependent demand is the demand that is created due to the bill of material (BOM). Supply planning and demand planning in SAP deals with three different types of demand:

1. *Initial Demand:* Forecast, Sales Order, etc.

2. *Dependent Demand:* Demand that is created by the explosion of the BOM.

3. *Distribution Demand:* Demand that is sent from one location to the next based upon how the locations are set up as sources of supply.

There are several settings that are on the PP/DS tab, which are actually just copied over from the SNP 2 tab. If one saves the fields on either the SNP or PP/DS tab, the information is saved on both tabs.

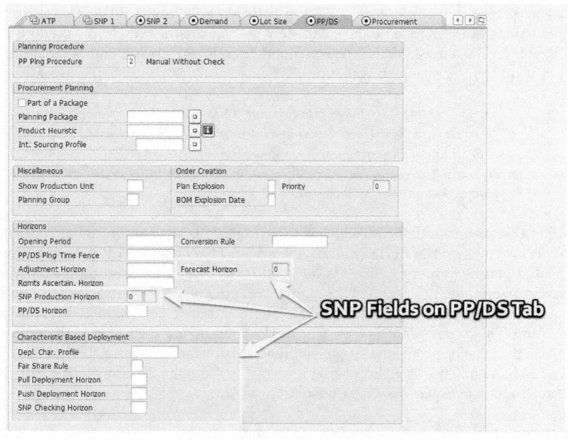

Notice the fields above on the PP/DS tab.

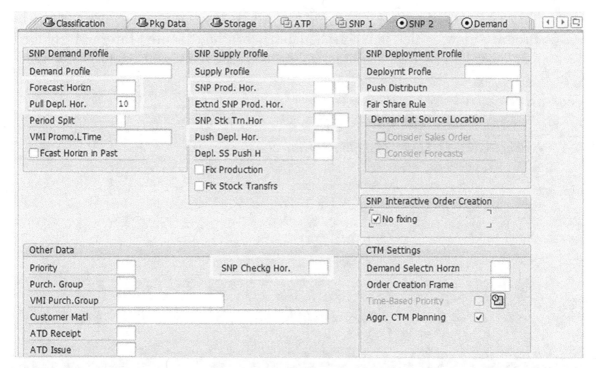

Above, we can see the same fields on the SNP tab. These fields are on the PP/DS tab because PP/DS can create unconfirmed stock transport requisitions (STRs) with an MRP run (which is a PP/DS heuristic—SAP_MRP_001). This is just one of many heuristics that are available within PP/DS. To see all the heuristics, view the following article: http://www.scmfocus.com/sapplanning/2008/09/21/ppds-and-snp-heuristics/. As of SCM 7.0, PP/DS can now also create confirmed STRs by performing deployment (although it is only meant for characteristic-based planning, which is almost never used).

The new PP/DS heuristic can consider the "characteristic values of the receipt, requirement element for allocating the quantity." This requires the following:

1. The Planning Area 9ASNP-PP/DS be initialized for the current planning version. Planning Areas hold the key figures and the key figure aggregates what the planning books depend upon and represent.

2. The heuristic SAP_DEPL_SNG, SAP_DEPL_MUL performs the deployment run for PP/DS and creates the confirmed STRs based upon the fields in the characteristic-based deployment area of the PP/DS tab.

3. The SAP_DEPL_SNG, SAP_DEPL_MUL heuristic is executed from the Product View, which is where heuristics can be executed interactively for a single product location combination, by selecting the variable heuristic options.

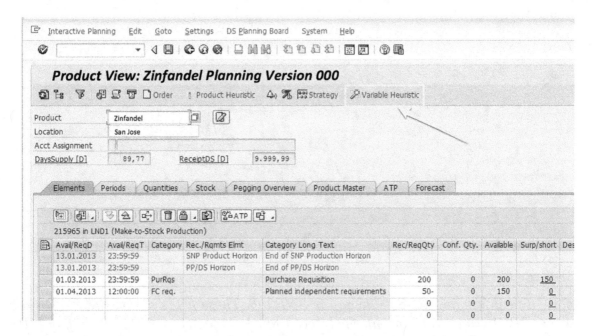

I could keep showing how this works, but I consider this functionality to be a waste of time. First, companies very rarely use characteristic-based planning (CBP) in APO. If a company wanted to use CBP, there are far better solutions, and I am not even convinced that APO CBP is sustainable. There are a number of areas of functionality in APO that were developed and are frequently dropped into sales presentations, but are not actually implementable. Second, by enabling PP/DS with supply planning functionality (but for an almost unused functionality), SAP development has further increased the complexity of PP/DS in a way that will benefit a vanishingly small number of clients. That the modules are increasingly overlapping into one another is a problem with APO. When PP/DS can perform supply planning activities and GATP can trigger production orders, the whole solution just becomes more confusing. However, I am torn, because if someone asks whether deployment can be performed in PP/DS, I have to tell them that it is possible, but then I have to get into the very minimal application of this new functionality. In fact PP/DS can also perform deployment, but some of the functionality that SAP development has included in APO doesn't make a lot of sense and won't be used by many clients.

With that, let's get back to the topic of demand types. The initial demand is easy enough to understand as it comes from either forecasts or sales orders. For a forecast or sales order to be processed at a location, it must be assigned to the location. The distribution demand is the demand between locations. Dependent demand is demand that is due to the explosion of a BOM, as shown below:

Dependent Demand

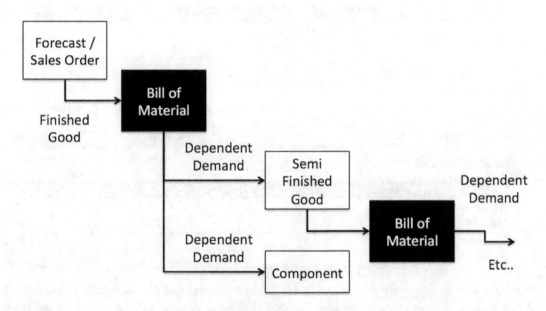

Both distribution demand and dependent demand are driven by the methods that are used to plan the supply and production network. All of these various demand types are shown as key figures in the SNP planning book.

Planning Book Mockup

	Key Figure Name	Common Order Categories	10/5/12	10/12/12	10/19/12
Demand	Forecast		5,000	1,000	300
	Sales Order		5,500	-	-
	Distribution Demand	Unconfirmed Purchase Requisition (either internal or external source of supply)			1,200
	Distribution Demand Confirmed	Confirmed Purchase Requisitions or Purchase Orders			
	Dependent Demand				
	Total Demand		**5,500**	**1,000**	**1,500**
Supply	Distribution Receipt	Stock Transport or Purchase Requisition			200
	Distribution Receipt Confirmed	Stock Transport Orders and Purchase Orders			
	In Transit		500		
	Production Planned		2,800	600	1,300
	Production Confirmed				
	Total Receipts		**3,300**	**600**	**1,500**
Stock	Stock on Hand		1,000	600	600
	Max Stock Level		600	600	600
	Reorder Point		500	500	500
	Safety Stock		500	300	300

In the above mock-up of the planning book, I have greyed out the key figures that match all of the demand types that are used in SNP. These comprise the demand section of the planning book (the planning book can have different key figures shown, hidden and added, so this mock-up will not represent every standard SNP planning book).

Another important limitation of capacity leveling, and one which is often left out of the comparison of the two approaches, is found in the following quote:

*Capacity leveling does not take dependent demands into account. Since leveling is only performed locally on a resource, it can lead to other resources being overloaded, additional on-hand stocks being created, or shortfall quantities being generated. — **SAP Help***

Dependent demand is something that is, of course, managed by constraint-based planning, which is another important distinction between the capabilities of the two approaches. If a higher-level resource is constrained and then distributes dependent demand along the BOM to another constrained resource, both resources are constrained automatically. However, in capacity leveling, the resources are co-managed, meaning that capacity leveling must be run in a sequence that is consistent with the flow of the BOM: higher level items (finished goods) first, and then semi-finished goods second, etc.

Capacity Leveling versus Constraint-Based Planning

		Requirement	Capacity Leveling	Constraint Based Planning	
Basic Functionality	1	Number of Steps in Process	2	1	
	2	Ability to Constrain Dependent Demand Resources	No	Yes	Where Most Companies Stop
Complex Functionality	3	Can Constrain Multiple Chained Resources in a Sequence	No	Yes	
	4	Can Bi Directionally Constrain Multiple Chained Resources in a Sequence	No	No	

I will discuss the more complex requirements, numbers 3 and 4, later in this chapter. However, this graphic should successfully explain that there is much more to the topic of constraint-based planning than whether a resource is or is not constrained.

On the other hand, in SAP, the capacity leveling heuristic can *"raise changeable SNP planned orders that use sub-resources of the resource to be leveled to the level of the resources to be leveled before it executes capacity leveling."* — ***SAP Help***

So SAP does address this with capacity leveling—if the setting is used—but does not do so as comprehensively as the SNP optimizer.

Backward and Forward Production Scheduling

Scheduling and scheduling direction is one of the most important features and settings in supply chain software that performs both capacity leveling and constraint-based planning. Backward and forward production scheduling has more implications than backward and forward consumption. However, it is much simpler to understand because, while the primary goal of backward and forward forecast consumption is to prevent over-ordering, (see the article for more details on forecast consumption at http://www.scmfocus.com/sapplanning/2011/06/09/forecast-consumption-setting/) backward and forward scheduling deals with "when" demand will be satisfied. Scheduling can be performed in two directions, but there are more than two options because the directions can be combined in one scheduling setting. Furthermore, different supply planning methods allow for different scheduling to be performed.

To understand how to use the different scheduling alternatives that are available in systems, it is important to begin with a definition of each of the scheduling types. SAP's definition of how the three scheduling options work in its system is as follows. These definitions apply for both SAP ERP and for SAP APO; however, other applications with supply and production planning functionality work very much the same.

1. *Forward Scheduling:* For the start date, the system uses the beginning of the period in which the production quantities were entered. From this start date, the system calculates in a forward direction to determine the finish date. The system displays the order quantities on the production start date.

2. *Backward Scheduling:* For the finish date, the system uses the end of the period in which the production quantities were entered. From this finish date, the system calculates in a backward direction to determine the start date.

3. *Backward / Forward Scheduling:* Here the system works in two steps:
   a. In the first step, the system uses the end of the period in which the production quantities were entered as the finish date. From this finish date, the system calculates in a backward direction to determine the start date.
   b. In the second step, the system uses the beginning of the period calculated in step one and then schedules forwards. Order processing commences at the beginning of the start period calculated by the system and ends in the period specified by the planner.

Some systems such as MRP, which is the supply planning method in SAP ERP, are run with backward, then forward scheduling by default. However, other supply planning methods, such as SAP CTM, do not have the ability to perform backward scheduling first and then forward scheduling, and can only perform either backward or forward scheduling in one planning run. Any supply planning system that is unconstrained must first perform backward scheduling. MRP is unconstrained and this is why there is no option to use only forward scheduling. SAP CTM has the ability to be constrained, and for this reason can be run with forward scheduling. In fact, with CTM, SAP does not offer the option to begin with backward scheduling and then move to forward scheduling, which is the default method of operation for MRP in SAP ERP.

More on backward and forward scheduling is available at the article below:

http://www.scmfocus.com/sapplanning/2012/06/27/backward-scheduling-forward-scheduling-sap-erp-sap-apo/

Backward scheduling is the most common scheduling direction. Backward scheduling works from the need date, and calculates the activities necessary to provide material availability "backwards" from the need date. Forward scheduling works much more simply, and schedules activities to take place as soon as possible, as if the material demand date is immediate. Performing forward scheduling only in a supply and initial production planning run (and no backward scheduling), "front loads" resources (also known as "fill to capacity"). When applied to internally produced items, machine and labor resources will be employed to product material prior to their need. One might observe that this is wasteful, but in fact the activity can be the correct action to take. Companies do not build capacity to match the peak demand periods throughout the year. Some companies will find themselves unable to meet capacity at certain times, meaning that they have the option of producing early, producing late (if the customer accepts late shipments), or denying the order. Furthermore, the setup involved in some products is significant when compared to the inventory carrying cost of keeping the material in stock. One company I worked with was able to produce a full year's demand for an item in three hours, but the setup time to produce this item was four hours. It would not make sense for them to break the single yearly production run into two 90-minute runs to avoid storing inventory.

Understanding the Primary Benefit of Forward Scheduling

Forward scheduling (also known as front loading) with internal production is primarily a trade-off between producing early and carrying inventory, or not producing early, and not filling one's production capacity. Forward scheduling allows a company to produce and procure before the system would ordinarily schedule (meaning more inventory is carried prior to the inventory being consumed). Because companies typically do not have unlimited factory capacity or material availability and also because some factories—repetitive manufacturing environments in particular—require long production runs in order to achieve their potential production efficiencies, the ability to forward schedule production orders and their associated purchase orders can be the correct approach to configure a company's supply and production planning systems. More detail on forward scheduling is available at the link below:

http://www.scmfocus.com/sapplanning/2012/06/13/front-loading-resources-in-sap-snp/

The forward scheduling setting can be seen in the screen shot on the next page, which shows SAP CTM in the Planning Strategies sub-tab of the CTM Profile.

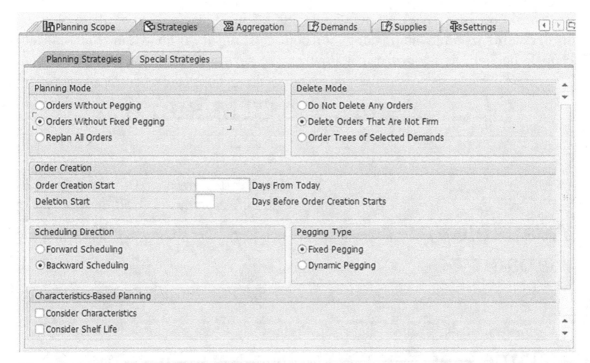

The options that are available with respect to backward and forward scheduling are very important for the output of the planning run. For instance, a procedure that begins with backward scheduling, and then switches to forward scheduling (which is the default for MRP in SAP ERP) will produce a very different output than one that begins with forward scheduling.

Scheduling Direction and Its Implications

As I have discussed, forward scheduling can be the correct setting in some circumstances. However, effectively leveraging APO to meet forward scheduling means more than simply changing the scheduling direction in the supply planning method. For instance, setting forward scheduling with CTM can lead to an interesting result, which is shown in the graphics on the next page. It can mean that higher priority customers, whose orders are run through the allocation supply planning procedure earlier, can consume a production resource sooner than they would ordinarily with backward scheduling. Forward scheduling plans all requirements as early as possible, without consideration for when the demand is actually needed, as is shown in following the series of graphics.

Before the CTM Run

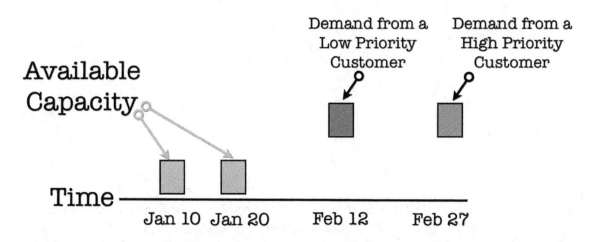

This is designed to show the state prior to the CTM run. Notice that there are two demands. The demand from the higher priority customer is further out in the planning horizon than the demand from the lower priority customer.

The CTM Run with Backward Scheduling

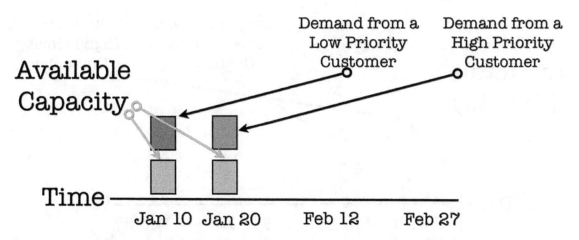

Under backward scheduling, as long as the demand from the higher priority customer can be met on time, the system will choose to peg or associate the demand with the later capacity.

The CTM Run with Forward Scheduling

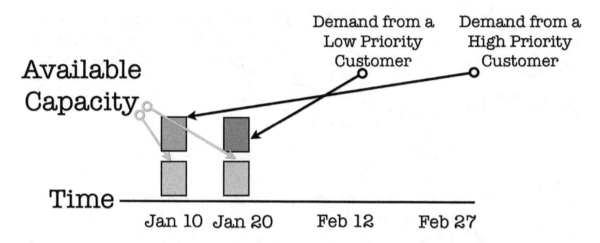

However, under forward scheduling, even if the demand from the higher priority customer can be met on time with the later capacity, CTM will assign the customer demand to the first available capacity. CTM does this because of the combination of CTM processing the higher priority customer prior to the low priority customer, along with the forward scheduling setting which plans all demands as soon as possible.

This is of course an undesirable outcome. However, there are several options, listed in the article below, to control this outcome and to prevent it from happening.

http://www.scmfocus.com/sapplanning/2012/06/27/backward-scheduling-forward-scheduling-sap-erp-sap-apo/

Scheduling Direction When Using Cost Optimization

The supply planning method of cost optimization can also be made to perform forward scheduling. However, most often it will not have a setting for scheduling direction. Instead, the scheduling direction is controlled by the costs that are set up in the optimizer, notably the storage costs. Storage costs make the model incur a cost for each day that a product is kept at a location. The inclusion of storage costs therefore creates an incentive in the model to delay production until that product is required. Therefore, the use of a storage cost promotes backward scheduling. If storage costs are set to "0," many optimizers will immediately switch to forward scheduling. When storage costs are included, the optimizer switches to backward scheduling as it has now been provided with an incentive not to minimize inventory.

Backward and Forward Scheduling with Procurement and Stock Transfer Planning in SNP

Forward scheduling can be used to create planned production orders, purchase requisitions, or even stock transport requisitions prior to when they would be scheduled under backward scheduling. Companies routinely pull forward their procurement and stock transports in anticipation of future demand. Usually forward scheduling procurement and stock transfers are performed manually rather than set up in the supply chain planning systems. In order for forward scheduling to work in a way that meets the business requirements (and not simply initiate all activities immediately), it is necessary to constrain the resources over which the activities are spread.

In addition to demand spikes, there can be other reasons for forward scheduling. For instance, if suppliers are unreliable, then forward scheduling could reduce the risk of non-delivery or late delivery by creating purchase requisitions as soon as possible. Another reason for forward scheduling could be if the price of a material is predicted to rise in the future. Both situations can arise when an industry becomes capacity-constrained. Years ago this happened in the aerospace industry with titanium, and therefore, any part made of titanium experienced lengthened lead times. In this situation, an alternative to forward scheduling is to adjust the lead times to make them represent the current environment.

As has been discussed throughout this book, it is common to plan with constrained production resources. It is much less common to plan with either supply planning resources (transportation resources, storage resources, etc.) or with the capacity constraints of suppliers (which would constrain the purchase requisitions).

The traditional output of a supply planning system is planned production orders—purchase requisitions and stock transfer requisitions. Planned production orders and purchase requisitions are created by the initial/network supply planning run, and stock transfer requisitions are created by the deployment run. (I will cover forward scheduling in the deployment run shortly.) If the initial/network supply planning run is set to forward scheduling, it will forward-schedule planned production orders and purchase requisitions unless controlled in some way.

The scheduling direction for each supply planning recommendation type (production, procurement, transfer) should be analyzed and determined separately. That is, just because a company wants to front-load its production schedule does not necessarily mean it also wants to front-load or forward-schedule its purchase requisitions (for materials that are not part of a manufactured product) or its stock transfers. Therefore, the company must determine which recommendations from the supply plan they wish to forward-schedule and which they do not wish to forward-schedule, and then adjust the settings in the supply planning application accordingly. When it comes to the initial/network supply planning run, the decision of what to forward-schedule is quite important, because there are two types of recommendations (planning production orders and purchase requisitions) that come from this planning run, and the same scheduling direction may not work for both. Manufacturing companies may create purchase requisitions for products that are part of a manufacturing BOM, and also for products that are purchased and then resold. When forward scheduling is enabled and there are production capacity constraints, the purchase requisitions that are part of manufacturing BOMs will only be brought forward to the degree that there is manufacturing capacity. However, purchased materials that are not part of a manufacturing BOM will simply be brought forward to as early as possible, unless the supplier's capacity is modeled and constrained. This is why it can make sense to place manufactured product along with the procured product that is an input to the manufactured product in one planning run and resold procured product in a separate run.

Forward Scheduling in the Deployment Planning Run: Push versus Pull Deployment

Stock transfers are not created with the initial or network supply planning run, but instead by the deployment run. (They are also created during the redeployment planning run, but as SNP does not perform redeployment I will not cover redeployment reports. Nor will I get into non-SAP applications as this book is focused on APO.) Information on redeployment can be found in the following article:

http://www.scmfocus.com/inventoryoptimizationmultiechelon/2011/10/redeployment/

Push and pull deployment is controlled in the following ways in SNP:

1. *By the Deployment Heuristic:* When controlled by either DRP or a common deployment heuristic, stock transfer requisitions can be based upon a push or pull deployment. When a condition at the receiving location (such as a demand or a target stocking level) controls the deployment, it is called a pull deployment. On the other hand, when the transfer is based upon moving material out from the sending location as quickly as possible, it is called a push deployment.

2. *By the Cost Optimizer:* A push or pull deployment can also be controlled by a differential in the storage costs between a sending and receiving location. A storage cost that is higher at the sending location rather than the receiving location creates a push deployment, as the SNP optimizer tries to minimize storage costs by moving stock out of the sending location before there is demand. The reverse situation causes a pull deployment.

I bring up the topic of push deployment because it is analogous to forward scheduling in the initial/network supply planning run. In a sense, stock transfers can be forward-scheduled through the use of push deployment, rather than pull deployment. This can be set up in SAP SNP in the SNP Deployment Profile, which is shown on the following page:

SNP Demand Profile		SNP Supply Profile		SNP Deployment Profile	
Demand Profile		Supply Profile		Deploymt Profile	
Forecast Horizn		SNP Prod. Hor.		Push Distributn	
Pull Depl. Hor.	10	Extnd SNP Prod. Hor.		Fair Share Rule	
Period Split		SNP Stk Trn.Hor		Demand at Source Location	
VMI Promo.LTime		Push Depl. Hor.		☐ Consider Sales Order	
☐ Fcast Horizn in Past		Depl. SS Push H		☐ Consider Forecasts	
		☐ Fix Production			
		☐ Fix Stock Transfrs			

Push Deployment Set Here

SNP Interactive Order Creation
☑ No fixing

Other Data				CTM Settings	
Priority		SNP Checkg Hor.		Demand Selectn Horzn	
Purch. Group				Order Creation Frame	
VMI Purch.Group				Time-Based Priority	☐
Customer Matl				Aggr. CTM Planning	☑
ATD Receipt					
ATD Issue					

Tabs: Classification | Pkg Data | Storage | ATP | SNP 1 | SNP 2 | Demand

Push deployment is set up for a number of reasons. The simplest reason is when there is only a small amount of storage space at a factory, which forces finished goods to be moved out immediately to a distribution center. However, push deployment, as with forward scheduling of the initial/network supply plan, can also be performed in anticipation of spikes in demand. Most companies know when these spikes in demand will occur, particularly seasonal spikes. Companies often do not build the capacity for their highest demand weeks or months into their supply network, and instead rely upon front loading or anticipatory stock transfers. Furthermore, a company can easily switch between push and pull deployment, as well as backward and forward scheduling, and therefore, adjust its supply planning strategy depending upon the time of the year.

Controlling the Application of Scheduling Direction to the Product-Location Combinations

Supply planning methods can be controlled for selective product-based scheduling. In most supply planning systems, different master data profiles can be created that allow the planning procedure to be applied to specific subsets of data. For instance, procured products or manufactured products that do not need to be forward scheduled would be placed in a separate profile with backward scheduling, or (first) backward (then) forward scheduling. All of this information should be maintained externally to the system in a product-location configuration database (which can be something as simple as a spreadsheet). This database allows analysts to compare and contrast more easily all of the settings that are applied to product-location combinations. This approach is described in the article below:

http://www.scmfocus.com/sapplanning/2012/07/04/product-location-database-segmentation-for-snp-and-supply-planning/

Forward Scheduling and Capacity Leveling

Forward scheduling can be used with an unconstrained supply planning procedure, but the unconstrained network supply plan must be processed with a capacity leveling procedure. The SNP capacity leveling heuristic can be run with forward scheduling as is shown in the screen shot on the following page.

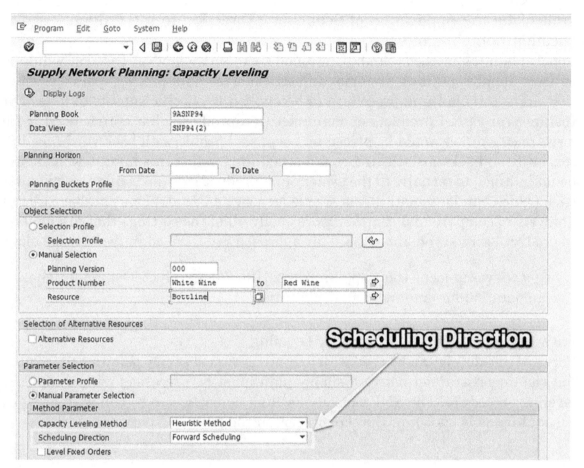

Capacity leveling is the second step when a non constraint-based method is used for supply or production planning. As can be seen from the screen shot above, SNP forward scheduling is an option, which I have selected.

Scheduling Direction and PP/DS

PP/DS has several forward scheduling heuristics that can be used to essentially do the same thing as has been described with forward scheduling in SNP.

1. SAP_PMAN_002—Infinite Forward Scheduling: Compact forward scheduling in the event of a scheduling delay in make-to-engineer or make-to-order production, based upon today's date or an entered date.

2. SAP_DS_01—Stable Forward Scheduling: Used to resolve planning-related interruptions using several BOM levels (infinite).

A full explanation of the heuristics available in PP/DS can be found in the article below:

http://www.scmfocus.com/sapplanning/2008/09/21/ppds-and-snp-heuristics/

However, the issue with forward scheduling in PP/DS is that these heuristics can only forward-schedule for the PP/DS Planning Horizon, which is typically no longer than a month. This is why forward scheduling requirements tend to fall onto SNP, as it has a far longer planning horizon.

Constraint-based Planning in the Applications

Having discussed constraints and constraint-based planning from a high level, now is a good time to show how the specific constraint-based planning settings look in SNP and PP/DS.

In SAP SNP, both the production constraints and the transportation constraints can be either continuous or discrete, as can be seen on both of the following screen shots.

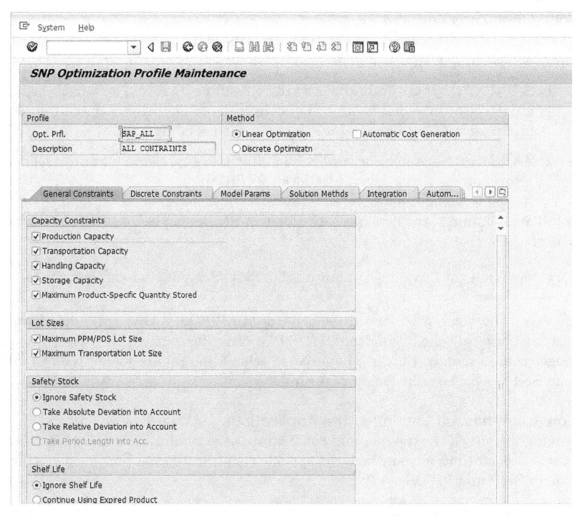

General or linear production constraints are activated or deactivated in this tab of the SNP optimizer.

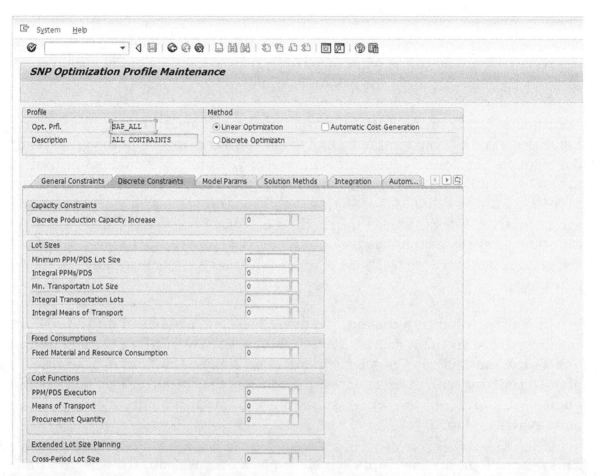

Discrete production constraints are activated or deactivated in this tab of the SNP optimizer. Once activated, the optimizer will incorporate the lot size located in the master data of the system. For instance, activating the Minimum PPM/PDS Lot Size will cause the system to respect the minimums of the resources that are a subcomponent of the PPM or PDS (PPMs and PDSs are SAP objects that hold routing, BOM, and resource information. To read more about these objects, see the article at http://www.scmfocus.com/ sapplanning/category/ppm-pds/). The minimum lot sizes can always be in the PPM or PDS, but this tab controls whether the optimizer respects these settings. Each lot size can differ per resource.

Continuous versus discrete optimization sounds more complex than it actually is. With both continuous and discrete forms of optimization, the optimizer looks at the resources to determine what is feasible; however, each form of optimization respects the constraints in the resources differently.

Continuous constraining (also known as linear or general) treats the resource capacities as if they can be started and stopped without any batching or lot sizing (for example, a resource could make 4 units, or 132 units; it would not be limited to batches of 50, 100, 200, etc.). Another restriction is that resources must be set to integer values. While it is true that supply planning generally occurs in integers, requiring integer values rather than accepting fractional values and then rounding after the fact makes the optimizer take more time. However, there are few options here because fractions of units don't make very much sense in supply planning.

When constraining in a discrete fashion, decisions are based on the specific capacity stated in the resource. Therefore, if a lot size of fifty is configured in a resource and "discrete" is selected, orders can only be created in quantities of fifty. Optimizing with discrete constraints makes the optimizer look at a greater number of limiting factors and results in a more accurate outcome, and thus a more realistic plan.

This overall topic of constraints is a bit complex because both the objective function and the constraints can be either linear or discrete. However, in this case I am referring to the linear or discrete nature of the constraints only.

Lot sizes can be activated for one category of lot sizes or for all of them. However, turning on discrete constraints for the optimizer makes the results less "optimal" (i.e., the objective function cost is higher) and causes the cost optimizer to take longer to solve. Specifics on this can be found at the link below:

http://www.scmfocus.com/sapplanning/2013/02/19/discrete-optimization-horizon/

I have found that clients can be confused on this topic, so it is worthwhile discussing it further.

Semi-Discrete Constraints

While often described in binary terms (i.e., "on" or "off"), discrete constraints can be enabled partially or completely, meaning that "discrete" can be enabled for some constraints but not for others. When describing discrete constraint optimization on a project, it is a good practice to be unambiguous about which constraints are discrete and which are continuous.

The Challenges of both Capacity Leveling and Constraint-based Planning

Companies tend to be very willing to buy software that can perform either constraint-based planning or capacity leveling. However they are much less willing to do the work necessary to make their capacity leveling and constraint-based planning projects a success. As I stated earlier, these systems can only provide output that is as good as the input they receive. Obtaining and updating resource information does take work, but what is the alternative? That is an easy question to answer: without good resource information, planned production orders are scheduled, which the factory cannot make, meaning more inventory and a less efficient use of production assets.

I realize that manufacturing is a dynamic environment with constant change, but most companies I have worked with have not done the basic work necessary to connect the centralized planning organization with the factories. This disconnect between the headquarters of companies and the factories is common, Companies worldwide suffer from poor integration between their factories and the central planning function. It goes beyond a lack of coordination, to the point where many of the personnel in headquarters and factories are contemptuous of one another. However, making this connection must become a priority; there really is no other way.

It's not as complicated or as difficult to bridge this gap as the low success ratio would indicate. Here are some common problems that I have run into with companies that are trying to improve their management of resources.

- Many companies do not maintain adequate databases on the history of work center and resource capacities or their sequence. This information is needed to set up the model correctly.

- Few companies are willing to make the effort to constantly update these resource constraints as things change in the supply chain.

- Resources taken out of commission for scheduled and unscheduled maintenance also need to be reflected in the resources capacity.

- Companies require time to improve both resource and BOM maintenance. Companies do not become experts at maintaining this type of information quickly, but require a cultural change that is focused on data maintenance.

- Companies often rely on spreadsheets for the maintenance of constraint information, but there is specialized software for this purpose that not only does the job better, but also allows more people with necessary information to participate in the process. Even though there are a number of BOM management system/software (BMMS) applications, most companies are stuck using antiquated approaches. I explain all of this as well as the best technology for managing the BOM in my book *The Bill of Materials in Excel, Planning, ERP, and PLM/BMMS Systems*. BOM management is still a challenging aspect of supply chain implementations in supply and production planning that relies on this data.

System Adjustments Required for Constraint-based Planning

While the differences between constraint-based planning and capacity leveling can seem like night and day, the same resources are used (and are mostly set up the same way) in SAP APO. In fact, I was asked this exact question by a previous client who was a bit intimidated by the prospect of moving to constraint-based planning. In my view, the changes required to move from capacity leveling to constraint-based planning are much more significant than the necessary technical adjustments.

Some individuals maintain that they don't have very good information on their resource capacities, but capacity leveling also requires accurate capacity information.

Without accurate information, capacity leveling will move demand to periods where capacity does not, in fact, exist.

The following major differences exist between the configuration of constraint-based planning and capacity leveling. The screen shots on the following page show the user interface views of constraint-based planning in APO:

1. *The Method Used to Perform Planning*:
   a. The SNP heuristic cannot constrain resources and must be capacity-leveled.
   b. Both CTM and the SNP optimizer can constrain based upon a resource.[3] Usually they are run this way, but they can also be run in unconstrained mode. Whether a resource is or is not constrained is set on both the resource and on the method. Secondly, not all resources need to be constrained or unconstrained. Each resource is set individually. CTM, for instance, can simply use the constraint setting set in the resource, so some resources could be constrained in a planning run, and others unconstrained.
   c. The rules for PP/DS are the same. Heuristics cannot constrain resources, but the PP/DS optimizer and CTM can perform constraining; again it is optional (CTM works for both SNP and PP/DS, but is mainly used in SNP).

2. *The Constraint Setting on the Resource*: This is shown toward the bottom of the following screen shot. Selecting the Finite Scheduling button means that, combined with the configuration of the supply or production planning method, the system is now performing constraint-based planning. However, both CTM and the SNP Optimizer can be set to ignore the finite scheduling selection on the resource.

3. *How the Resources Behave:* In the capacity planning book, when resource constraints are not used, then time buckets can have over one hundred percent applied to them. In these cases the time buckets will show as red. However, when constraint-based planning is used, the time buckets will always contain one hundred percent or less. Furthermore, capacity leveling

---

[3] Too often CTM and the SNP optimizer are described as "constraint-based methods." This is not the most accurate way to describe these methods. They are both methods that have the ability to constrain on resources; however, they can also be run unconstrained or even partially constrained.

can be run from within the capacity planning book, which is called running capacity leveling interactively. This option allows the user to select the number of periods—or the planning horizon—for the capacity leveling and runs capacity leveling only for the selection.

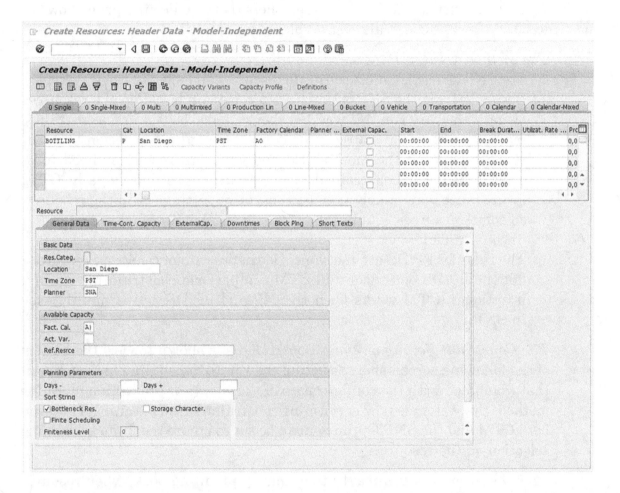

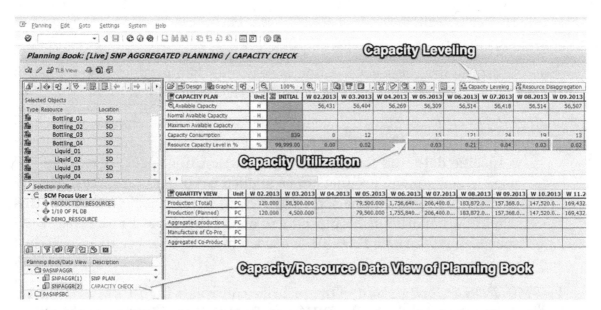

Here the SNPAGGR(2) data view (the view for aggregated planning) must be selected, which puts the planning book into a resource view. I can then select all of the production resources in the system, and open them in the spreadsheet to the right. The capacity utilization percentage shows in the final key figure toward the bottom of the first spreadsheet to the right. If the resources have been constrained, then the leveling is managed in the initial run. If the SNP Heuristic is used or the SNP Optimizer is unconstrained, then one can perform capacity leveling interactively for the resources selected in the Selection Profile to the left. Capacity leveling can also be performed from the capacity leveling transaction.

Priorities Used by Constraints in Both Supply and Production Planning

Listed below are the different types of priority fields available in SNP and PP/DS:

1. *Product Priority Field:* Located on the SNP 2 tab of the Product Location Master. This field is related to the following areas of functionality:
    a. *Capacity Leveling:* Whether to capacity-level based upon the product priority is defined in the SNP capacity leveling profile. This field is called "Order Prioritization," as described in the section in this book entitled, "How Capacity Leveling Works in Practice." Prioritization is one option that can be used. The options available include the following:
       - Prioritization
       - Without Priority

- (by) Order Size
- (by) Product Priority

(If the first or second option is selected, then product priority set in the Product Master does not affect the capacity leveling).

   b. *PP/DS Order Sequence:* It controls the order by which material is scheduled in PP/DS. The priority helps to sequence all production orders for a specific date and for a specific production line (priority can only be used to sort among products to be applied to similar resources). Higher priority products will be scheduled before lower priority products (this would only matter if both products share the same PPM/PDS) if the scheduling heuristic SAP001 is used. This heuristic is called "schedule sequence based upon the priority."[4]

2. *PPM/PDS Priority Field:* This priority controls which PPM or PDS for production is used first. The PPM and PDS is a combination of the BOM, routing, and resource. So a company can set up multiple versions of the PPM or PDS, which differ across the dimensions of the BOM, routing and resource (or all three) and then set the preferred PPM/PDS. When the capacity or the material of the primary PPM/PDS is consumed, SNP will automatically move to the next PPM/PDS in the priority sequence. (More on the PPM/PDS at http://www.scmfocus.com/sapplanning/2012/07/27/the-connection-between-boms-routings-work-centers-in-erp-and-ppms-pdss-in-apo/.)[5] The PPM/PDS priority does not help SNP make decisions between locations. Instead, it helps SNP and PP/DS choose the best PPM/PDS from a series of alternatives within a location.

---

[4] I should say that I am unconvinced of the idea of the production planning and scheduling system determining the priority of products to be produced. To me, this is the natural role of the supply planning system. Therefore, in my view the priorities should have been taken care of before the planned orders arrive at the production location.

[5] One might ask the question as to how switching to a different PPM/PDS based upon a material constraint would make any difference. It's a good question. An identical finished good can have both different input materials as part of its BOM (remember a different BOM means a different PPM/PDS), but different machinery can also mean a different consumption level of material. Some machines (resources) are inherently more efficient and waste less product than others. Therefore, one PPM/PDS could be material-constrained in meeting demand, whereas another PPM/PDS may not be. This is part of the larger point that the alternate PPM/PDS can vary among any dimension of BOM, resource, or routing.

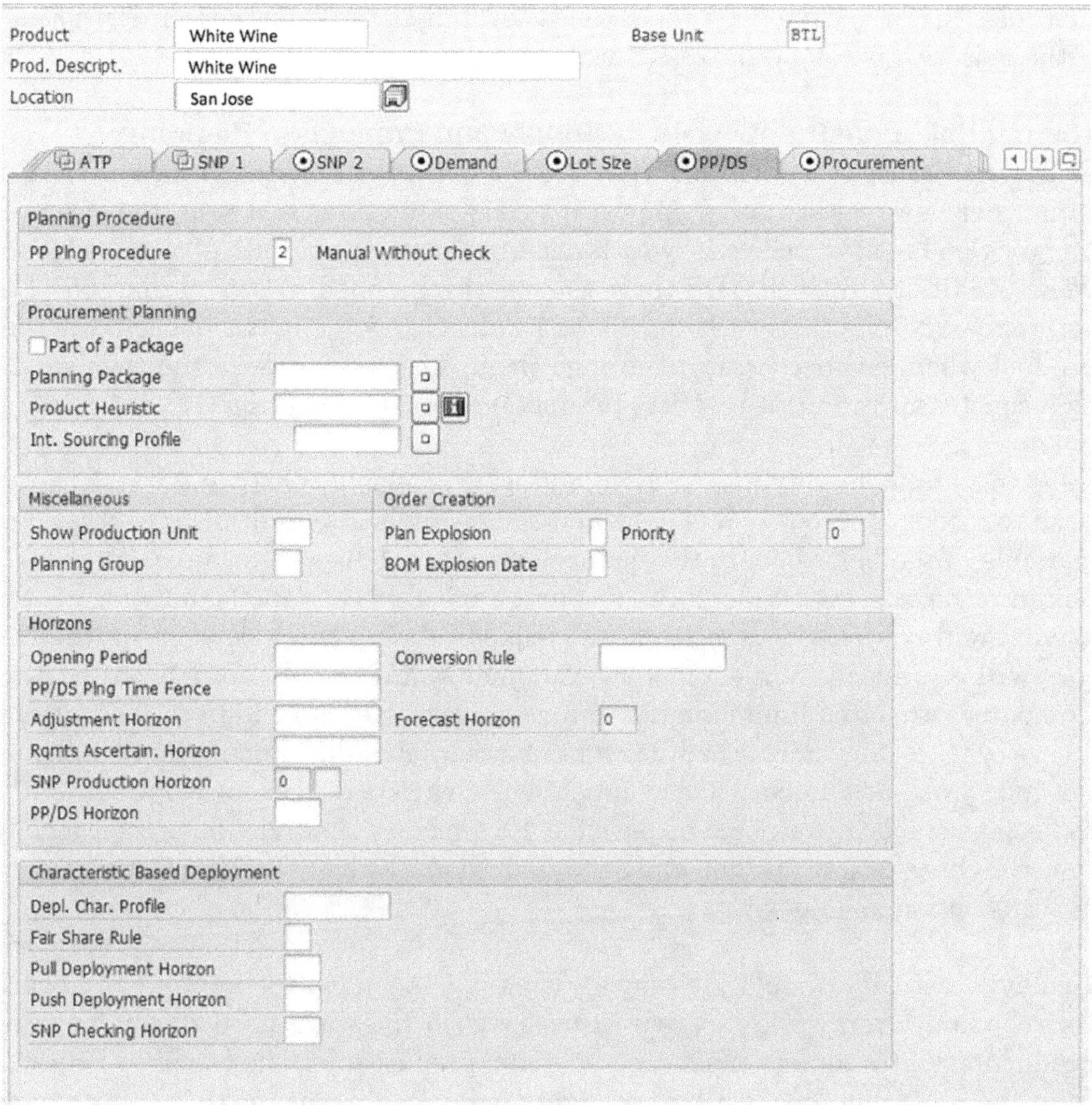

Product	White Wine		Base Unit	BTL
Prod. Descript.	White Wine			
Location	San Jose			

ATP   SNP 1   ⊙SNP 2   ⊙Demand   ⊙Lot Size   ⊙PP/DS   ⊙Procurement   ◀ ▶ ▢

Planning Procedure

PP Plng Procedure   [2]   Manual Without Check

Procurement Planning

☐ Part of a Package
Planning Package
Product Heuristic
Int. Sourcing Profile

Miscellaneous		Order Creation		
Show Production Unit		Plan Explosion	Priority	0
Planning Group		BOM Explosion Date		

Horizons

Opening Period		Conversion Rule	
PP/DS Plng Time Fence			
Adjustment Horizon		Forecast Horizon	0
Rqmts Ascertain. Horizon			
SNP Production Horizon	0		
PP/DS Horizon			

Characteristic Based Deployment

Depl. Char. Profile
Fair Share Rule
Pull Deployment Horizon
Push Deployment Horizon
SNP Checking Horizon

The Product Priority Field can be entered not only on the SNP 2 tab, but also the PP/DS tab. However, it should be noted that priorities do come at a cost. Specifically, it means that the lowest cost is no longer the only objective, nor is it the only change that is required when one desires to include priorities in the system. For instance, SAP recommends that capacity leveling not take into account priorities if the objective is to maximize the utilization of resources.

The Multiple Constraint Types in Supply and Production Planning

While we discuss resources as if they are the main focus of constraint-based planning, there are several other important constraints. For instance, the resource has a calendar assigned to it, which declares when the resource is available for work. Additionally, in PP/DS there is something called a "set-up matrix." The set-up matrix tells the system how long it takes to switch between each different product when it is necessary to change them. The set-up downtime then serves as a constraint in that it restricts production for that downtime.

Labor pool functionality in SAP ERP and APO is a labor constraint, which can be used to model the labor in a factory and can be applied to multiple resources. For example, if the labor pool in the factory can only produce eight hundred widgets on ten resources, even though the resource capacity is one thousand widgets, the system will not accept more work and some of the resource capacity in the factory will go unused. While a factory could add workers, they often choose not to. Companies are often limited in how many workers they can add in the short term. Therefore a factory may schedule eighty workers in a day, and cannot (or choose not to) grow that number by very much. For a variety of reasons, companies like to keep as stable a workforce as possible. So a factory that may have the capacity to use eighty workers one day and one hundred and twenty workers the next, may see this option as undesirable.

In a hypothetical example, let's say a company has ten production lines, and each line consumes ten workers when in use. Which lines are used throughout the production shifts change depending upon the product mix. The company wants to limit the consumption of workers to eighty per day. This means that the company does not want production orders to be scheduled for any day that would push the number of workers over eighty. Correspondingly, on average, only eight machines are utilized at any one time (although, again based upon the product mix, all machines may be used throughout the day, or nine of the ten machines may be used at some point during the day, but on average the total machine utilization is eighty percent). This is where labor pools come into play. A labor pool is itself a resource which represents labor, but which is assigned to a resource or multiple resources. Labor pools can be assigned to resources that are highly automated, or a resource can be a packing area where the worker is really doing most of the work unassisted. However, the creation of the resource is the standard design, with the labor pool being assigned to that resource.

Labor Pools in SNP versus PP/DS

A labor pool is a constraint for labor, and a co-constraint (along with resources) that accounts for the availability of labor within a facility. The labor pool is modeled as a capacity category and is associated with the machine work center for which it is needed. Once in APO, it appears as a *multi-mixed resource.* The routing configuration is created to account for labor requirements on the machine. Once these routings are moved to APO, they appear as a *secondary resource*. Both the labor resource and machine resource that the labor works appears in the detailed scheduling (DS) board for the production planner. The production planning can then perform the drag-and-drop scheduling on the DS board, thereby moving the operations and orders from one time bucket to another. This results in automatic movement of orders onto other machine resources once the labor resource is being scheduled. The article below provides more information on labor pools as a constraint.

http://www.scmfocus.com/sapplanning/2012/08/08/labor-pool-as-a-constraint-in-snp-and-ppds/

In most cases we discuss the machine capacity as the primary driver of the constraints for production. However, things like calendars, set-up times, and labor constraints are just as important. Using labor pools is a way to place an extra, or co-constrain a set, of resources that represents the labor consumption of a resource for its operation. While many factories are quite automated, labor is still required at some point in the process. Adding labor pools makes constraint-based planning more realistic and also allows the company to set the maximum amount of labor they want to work for a time period, in addition to ensuring that the plan will not create planned orders that require it to scramble to add more labor.

Most companies do not have perfectly matched production and demand. Therefore, a strategy for keeping a high utilization rate of labor and resources is to forward-build the necessary inventory during valleys in demand. This is called "front loading" or "forward scheduling" and is described in the article below:

http://www.scmfocus.com/sapplanning/2012/06/13/front-loading-resources-in-sap-snp/

Labor pools can be applied to any number of resources, creating a multi-resource constraint that will stop production from being planned before the actual resources are maxed out. Labor pools are also an excellent tool to create an integrated constraint.

Setting Up Labor Pools

Labor pools are created in R/3, and then brought across to APO. Labor pool resources can be created and assigned to a work center in SAP ERP. Below is a routing that connects two work centers. These are "TESTW1" and "TESTW2." You can see that there is labor assigned to each.

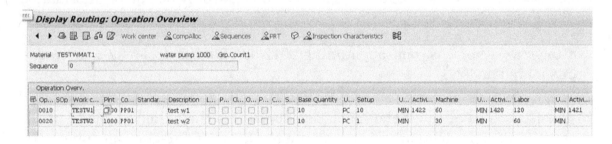

Labor Pools in SAP ERP

Labor pools are set up in SAP ERP. In the following graphic, we can see the labor pool "T-Pool" as it is assigned to the Work Center TESTW1.

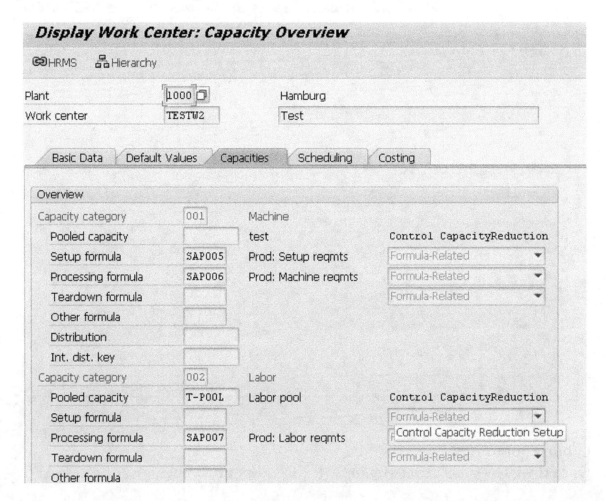

Then, in the MRP4 tab, one can see the Production Version by selecting the Production Version button in the middle of the screen.

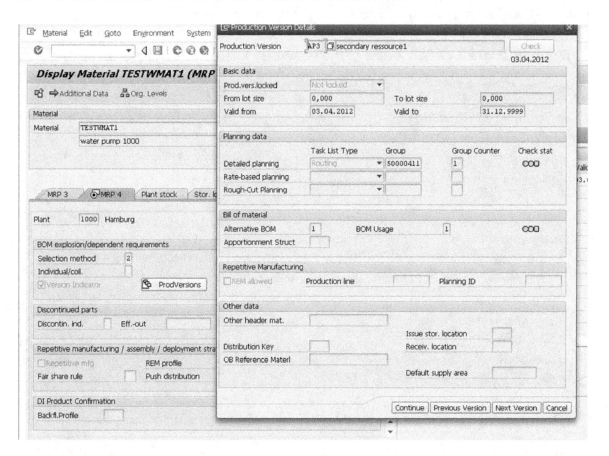

Labor Pool in APO
Once the labor pools are set up in SAP ERP, they are brought over to SAP APO.

Display Work Center: Capacity Overview

⊙ HRMS   ⊞ Hierarchy

Plant   1000 ⊡   Hamburg
Work center   TESTW2   Test

| Basic Data | Default Values | Capacities | Scheduling | Costing |

Overview

Capacity category	001	Machine	
Pooled capacity		test	Control CapacityReduction
Setup formula	SAP005	Prod: Setup reqmts	Formula-Related ▼
Processing formula	SAP006	Prod: Machine reqmts	Formula-Related ▼
Teardown formula			Formula-Related ▼
Other formula			
Distribution			
Int. dist. key			
Capacity category	002	Labor	
Pooled capacity	T-POOL	Labor pool	Control CapacityReduction
Setup formula			Formula-Related ▼
Processing formula	SAP007	Prod: Labor reqmts	Control Capacity Reduction Setup
Teardown formula			Formula-Related ▼
Other formula			
Distribution			
Int. dist. key			

Here we can see that the PDS with the material TESTWMAT1 is in the PDS. TESTWMAT1 is assigned to the labor pool T-Pool.

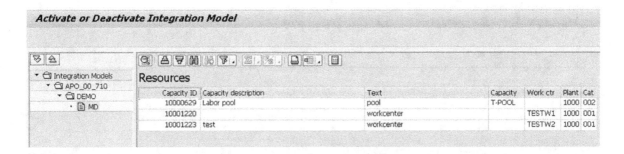

By assigning a labor pool to a resource, the constraint becomes essentially a "co-constraint." That is, the resource will stop scheduling production when either the machine or the labor pool is consumed. The nice thing about the labor pool is that it can be assigned to multiple resources. Once the specific labor pool—in this case the "T-Pool"—is consumed, that resource is no longer available to schedule. In fact, there could be one labor pool for all the resources, which would be the simplest way of modeling the labor. Alternatively, if the labor is more specialized (which is more likely the case), multiple labor pools can be created and then assigned to different resources.

Notice that the T-Pool_1000_002 shows up as a resource in the Capacity Requirements screen, along with the the resource that is shown in the PDS below.

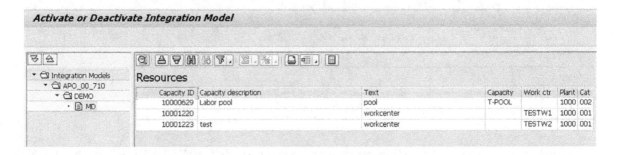

Notice in the following graphic that the labor pool shows up in the the Detailed Scheduling Board. This is the main scheduling view in the PP/DS system.

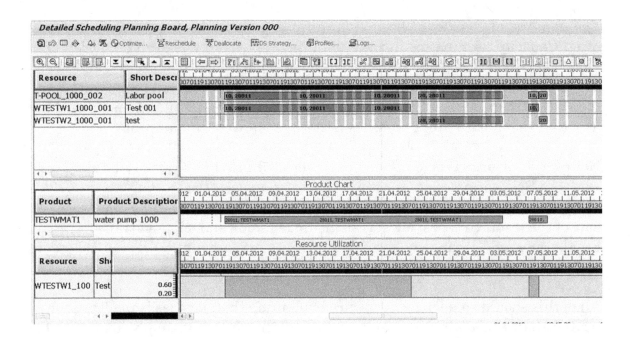

Long-term Constraints

In this chapter, the entire discussion of constraints has centered on respecting constraints. However, another supply planning run is designed to help the company determine which constraints are worthy of investment so that they are no longer constraints. This is a sales and operations planning (S&OP) run, and it is a special planning run dedicated to understanding the investment needs of operations and helping Finance determine the best places for their investments, rather than focusing on moving demand forward or backward in time to match capacity and availability of materials.

Conclusion

Constraint-based planning (or finite planning) works by setting up limiting factors in a model. Resources can be anything from factory equipment to transportation units to handling equipment. When resources become filled, constraint-based planning moves the activity to a different period (and sometimes a different resource). Assuming the resources represent reality, and the master data on the limiting factors in the model are accurate, a plan generated in this way is feasible and implementable.

The opposite of constraint-based planning is capacity leveling or infinite planning. Capacity leveling (or infinite planning) allows any load to be placed onto any time period. Capacity leveling is applied in a second step to smooth out the peaks and valleys. Infinite or unconstrained planning followed by capacity leveling was introduced in software before constraint-based planning and is still the way that the majority of companies perform supply chain planning.

Not every resource type can be capacity-leveled in APO. Capacity leveling only works in SNP for production resources and transportation resources, ***and does not work for storage or handling resources***. Capacity leveling only works on four of the resource types: bucket, single-mixed, multi-mixed, and transportation. Furthermore, SNP capacity leveling only takes into account SNP planned orders and SNP stock transfers.

All of this should go a long way toward explaining that, besides being less sophisticated than constraint-based planning, capacity leveling simply applies to far fewer possible entities than constraint-based planning in SAP APO. Many of the differences between capacity leveling and constraint-based planning tend to be overlooked on projects. It's as if many of the extra features of constraint-based planning—such as the ability to constrain through the entire BOM in one step— get lost in the equation.

The two approaches are also compared without considering the extra work required to perform capacity leveling. It's assumed that the company will simply hire more planners or have the existing planners pick up any extra work required to perform the more manual planning. However, almost every client I have worked with has fewer planners than what they need for the existing work level, and I don't see that changing in the future. Typically, a company has the absolute fewest number of planners, with the absolute lowest level of training, to get the job done at a very basic level.

Both constraint-based planning and capacity leveling work off of backward and forward scheduling; that is, how the overloads are moved backward and forward in time is controlled by configuration. For instance, a company may only want to backward-schedule five days, but forward-schedule ten days. The scheduling alternatives vary depending upon whether the supply plan (and initial production plan) is created in a one-step process (that is capacity constrained) or a two-step process, and these options also vary depending upon the system and the method used. For instance, in SAP ERP, the default scheduling method for MRP is (first) backward,(then) forward scheduling. The scheduling direction is set on the method in SNP and PP/DS, and the number of days forward or backward is set at the product-location combination, meaning that the number of days in either direction can be different per product location.

There are many ways to configure the SNP Optimizer Profile, and all of the tabs in the SNP Optimizer Profile were shown and explained in this chapter. Additionally, aside from the system settings that are explicit within the SNP Optimizer Profile, we covered the other system adjustments that are necessary for using constraint-based planning.

The software capabilities of resource management, capacity leveling and con-straint-based planning of SAP APO, as well as the capabilities of every best-of-breed vendor I have been exposed to in supply and production planning, are now so sophisticated that the primary problem is not software and not hardware, but the level of communication and information exchange between those that maintain the planning system and those that have the knowledge regarding the resource constraints in the factories. The management at many companies does not seem to grasp how bad the situation often is. On one factory visit I was asked if I came from the "Puzzle Palace." I asked the factory manager what he meant by that and he told me, "I call it that because no one there knows what they are doing." I have had this same "vibe" or similar indicators at almost every project I have worked on in supply and production planning. Most often, factory management do not think that the individuals in headquarters necessarily lack in intelligence, but that they are completely out of touch with what is happen-ing in the factory. For instance, companies often discuss how beneficial it would be if they could only model the constraints of their suppliers. However, this is a strange statement given that these same companies cannot even accurately model the constraints of their internal factories. To take advantage of all that the sophisticated software has to offer for managing resources, this situation will have to improve in the future.

Methods that perform constraint-based planning are often described as "constraint-based methods," the implication being that they are exclusively used for constraint-based planning. However, the truth is much greyer than that. Constraint-based methods can perform constraint-based planning but they are not always set up this way. Secondly, any of the constraint-based methods can be used with some resources that are constrained and some that are not. This is an important distinction because it's quite common to not only have mixed constrained resources, but resources that are constrained by the active or live version (the version that interacts with the ERP system.) See this article for more details:

http://www.scmfocus.com/sapplanning/2013/01/31/version-copy-in-apo/

So, the best way of describing the constraint-based methods is that they are **capable** of constraint-based planning, not that they *are* constraint-based planning methods.

While resources tend to be the main area of focus when discussing constraints, in fact there are several others, including the set-up matrix, which determines how long resources will be idle while a resource is prepped to create a new product. Another constraint type is the calendar that is assigned to the resource and determines when the resource is available to perform work. Labor is also another important and often overlooked constraint. Labor pool functionality exists in PP/DS, but not SNP. The labor pool is a "co-constraint" in that it is assigned to a group of resources; if the pool of labor becomes consumed before the resources, no more work is accepted. Labor pools must be created in R/3 and then brought over to APO.

While cost optimization has the objective function of cost minimization, both the optimizer in SNP and PP/DS can also consider priorities. The prioritization fields were described in this chapter. Priorities control which products should be produced over other products as well as which PPM/PDSs to use first before moving to lower-priority PPM/PDSs.

There are different planning threads that support different types of analysis. Most of this book is concerned with the initial supply plan as well as the production plan and the production schedule. These are the "live" planning threads, and the recommendations from these threads are eventually sent to SAP ERP for execution. However, there are other threads that are most often run in simulation versions of APO and that are used for different purposes. Two of these threads are the S&OP thread and the rough cut capacity planning thread. Rather using these two threads to create a feasible plan, they are designed to allow a company to determine where over-capacity occurs, and make adjustments to capacity that can either alleviate these constraints in the future or make broader changes than are typically taken with the active planning threads.

CHAPTER 3

Integrating Supply and Production Software with Constraints

Supply planning and production planning are most often discussed as separate activities. However, production constraints are both quite popular and is the most common type of constraint to place in the supply planning system. One of the benefits of using production constraints is that the supply plan is able to send a feasible production plan to the production planning application for planning and scheduling. The general objective is to pass a feasible plan; that is, the production facility has sufficient capacity to produce all of the planned production orders per week. If possible, the resources (labor and machine) should also be well-utilized. This can mean pulling forward production—sometimes a number of weeks in advance of needs—in order to prevent overloads. This is something that supply planning is able to do because it has a very long planning horizon (often twelve months), while the production planning horizon is generally between two and four weeks.

Focus On  Customizing the Production Planning Horizon Per Factory

The production horizon can be set differently per factory. The two settings that apply are as follows:

1. The PP/DS Planning Horizon

2. The SNP Production Horizon

Because both of these settings can be set at the Product Location Master in APO, they can be customized not only per location, but also per product-location combination.

First Draft Horizons

Location Type / Location Name	DC/RDC Europe	RDC Japan	RDC Singapore	RDC US - West	RDC US - East	RDC China	Factory US	Factory Germany	Factory Norway	Time Measurement
Demand Planning Horizon	21	21	21	21	21	21	21	21	21	Months
GATP Order Checking Horizon	14	14	14	21	21	14	14	14	14	Days
SNP Forecast Horizon										
SNP Planning Horizon (Weekly Run) (constrained at the finished good for unconstrained)	18	18	18	18	18	18	18	18	18	Months
SNP Planning Horizon (Daily Run)	12	12	12	12	12	12	12	12	12	Months
CTM Time Stream	12	12	12	12	12	12	12	12	12	Months
CTM Demand Selection Horizon	We can flexibly assign per product.									
Pegging Horizon	12	12	12	12	12	12	12	12	12	Months
SNP Production Horizon						21	14	7	7	Days
PP/DS Planning Horizon						28	21	14	14	Days
PP/DS Planning Time Fence						1	1	1	1	Days
PP/DS Adjustment Horizon										
Deployment Horizon	12	12	12	12	12	7	7	7	7	Days

These settings control the "hand-off" between the supply planning application and the production planning application. The longer the SNP Production Horizon, the more control the production planners have and vice versa. Both SNP and PP/DS can create planned production orders; however, the horizons control the following:

1. When only SNP can create planned production orders

2. When only PP/DS can create planned production orders

3. When both SNP and PP/DS can create planned production orders. (This is not mandatory. If the SNP Production Horizon is set to three weeks and the PP/DS Planning Horizon is set to three weeks, then there is no overlap. However, on most projects, there is an overlap between SNP and PP/DS, so that both systems for a time are creating and adjusting planned production orders.)

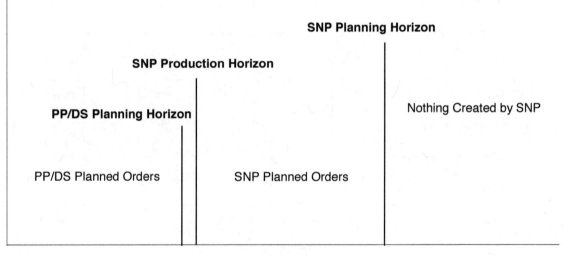

The SNP Production Horizon and
The SNP and PP/DS Planning Horizons

Timeline

SNP cannot create planned production orders inside of the SNP Production Horizon.

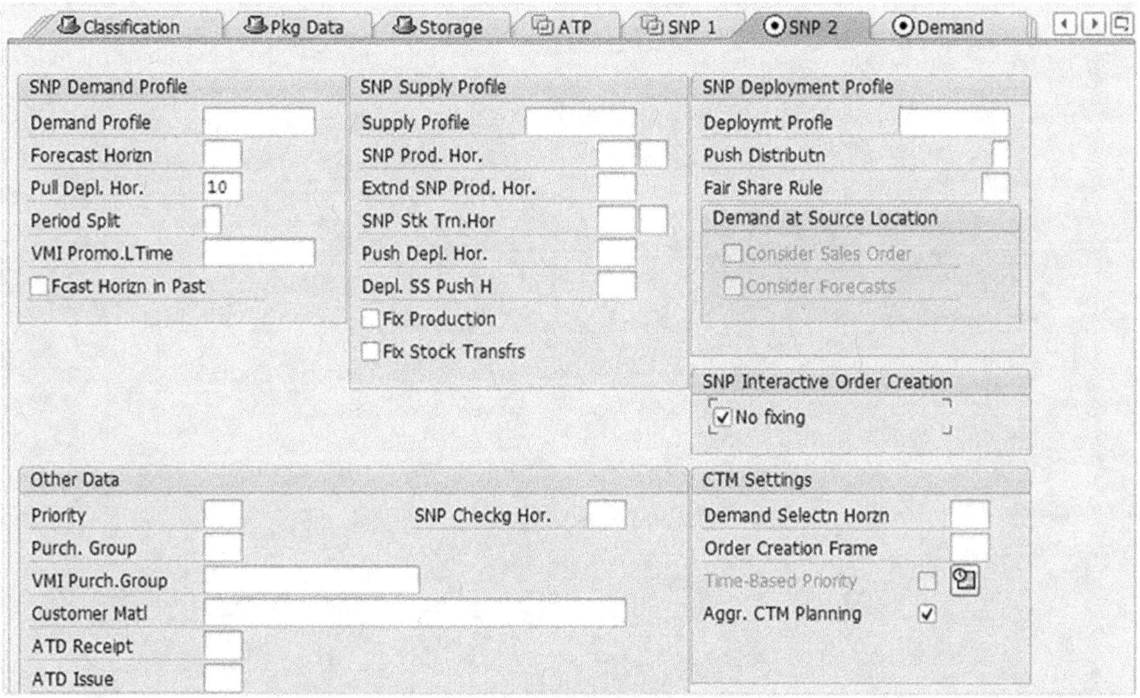

Classification	Pkg Data	Storage	ATP	SNP 1	⊙ SNP 2	⊙ Demand	◄ ► ⊡

SNP Demand Profile

Demand Profile	
Forecast Horizn	
Pull Depl. Hor.	10
Period Split	
VMI Promo.LTime	
☐ Fcast Horizn in Past	

SNP Supply Profile

Supply Profile	
SNP Prod. Hor.	
Extnd SNP Prod. Hor.	
SNP Stk Trn.Hor	
Push Depl. Hor.	
Depl. SS Push H	
☐ Fix Production	
☐ Fix Stock Transfrs	

SNP Deployment Profile

Deploymt Profle	
Push Distributn	
Fair Share Rule	

Demand at Source Location

☐ Consider Sales Order
☐ Consider Forecasts

SNP Interactive Order Creation

☑ No fixing

Other Data

Priority		SNP Checkg Hor.	
Purch. Group			
VMI Purch.Group			
Customer Matl			
ATD Receipt			
ATD Issue			

CTM Settings

Demand Selectn Horzn	
Order Creation Frame	
Time-Based Priority	☐
Aggr. CTM Planning	☑

The SNP Production Horizon is set on the SNP 2 tab of the Product Location Master.

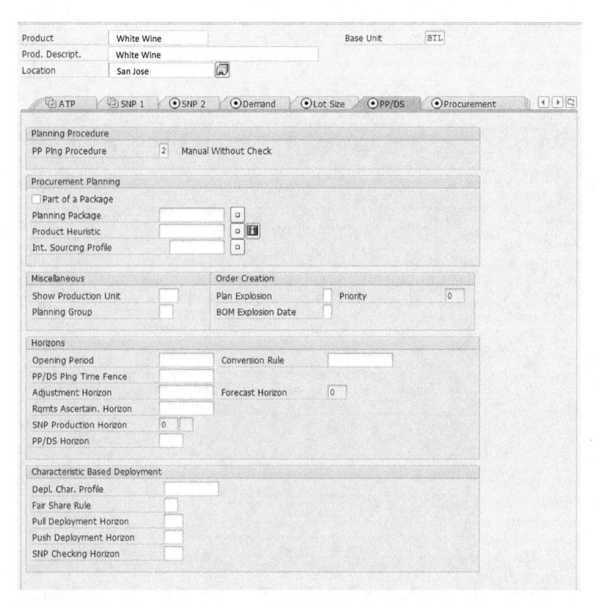

The PP/DS Planning Horizon is set on the PP/DS tab of the Product Location Master.

Integrating the Timings of SNP and PP/DS

In terms of the integration of timings between the different modules, of all the projects I have worked on, the connections between SNP and PP/DS draw the most interest and are in fact the most complicated. As I have discussed, GATP can be connected to PP/DS. In the vast majority of cases, SNP serves as the centerpiece module for planning connecting to DP, to GATP, and to PP/DS. There is nothing particularly unique about this design with regards to APO. Most of the other suites and individual non-SAP products work the same way. So we will now cover the most important time settings that allow SNP and PP/DS to work smoothly together.

SNP Production Horizon

The SNP Production Horizon is the horizon that divides the responsibility, in terms of time horizon, between SNP and PP/DS. Because SNP's most common implementation design is with production resources, a frequent question on APO projects is, "Where does SNP stop and where does PP/DS end?" There are several dimensions to how SNP and PP/DS differ with respect to planning production. As discussed in the book *Planning Horizons, Timings and Calendars in SAP APO,* one dimension is the time orientation of each module that SNP and PP/DS each work with. SNP stops at the time granularity of the day, while PP/DS goes down to the hour. By doing so, PP/DS gets the information it needs to perform detailed scheduling, which is one half of the focus of PP/DS. Another dimension is how far out along the planning horizon each system plans. The SNP Planning Horizon is typically set to a year, while PP/DS is typically set to between two and four weeks. As a result, SNP will always include the PP/DS horizon. However, the SNP Production Horizon essentially disables SNP production planning within the SNP Production Horizon. SNP planned orders only begin on the first day outside the SNP Production Horizon, so if the SNP Production Horizon is set for three weeks, SNP does not create planned orders until after the third week. Most often the PP/DS Planning Horizon is within this SNP Production Horizon. However, on most projects, the PP/DS Planning Horizon intrudes for a week into the SNP Planning Horizon. So if the PP/DS Planning Horizon were set to four weeks in this scenario, there would be a one-week overlap where both SNP and PP/DS can create a planned order. This can be confusing, because I have said that SNP will always be longer than the PP/DS Planning Horizon. Wouldn't the SNP Planning Horizon completely envelop the PP/DS Planning Horizon? The answer is that it does, but the SNP Production Horizon prevents SNP from creating planned orders within the SNP Production Horizon. Getting back to the topic of the overlap, it is ordinarily not a problem. In many cases SNP will have processed the time horizon before it gets to PP/DS for that week of overlap, so essentially PP/DS will come and process the overlapped portion of the planning horizon again. However, if planned orders have already been created, then PP/DS will have no need to create additional planned orders.

Extended SNP Production Horizon

The extended SNP Production Horizon allows manually created planned orders to be created within it, but prevents an SNP planning run from creating planned orders within the horizon. How the Extended SNP Production Horizon works is shown in the graphic below.

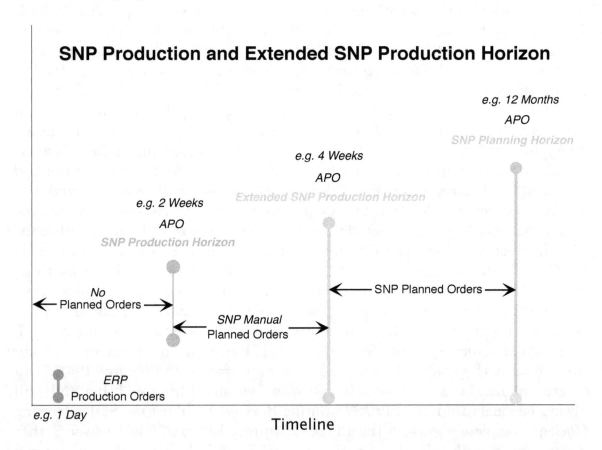

Understanding the Differences Between the SNP and PP/DS Optimizers

The following quote is from the book *Production Planning in SAP APO*.

> *The PP/DS Optimizer has specific properties that distinguish it significantly from the calculation procedure of heuristics and from the SNP Optimizer. Generally, the SNP Optimizer handles aggregates, period-oriented plans that are created and improved based on lean master data. Linear and analytical problem-solving procedures can usually meet such goals. But the situation looks much different when it involves the solution of complex problems from APO-PP/DS, problems that require consideration of several constraints. The effort needed to find an exact solution to the problem increases with the number of possible permutations. To put it more exactly, the effort needed increases by the factorial of the orders to be planned. Therefore, the PP/DS Optimizer uses selective procedures that lead to realistic computing times—but without a guarantee that the absolute optimal solution has been found.*

So, the PP/DS Optimizer has more complexity to deal with each location, and is more detailed than the SNP Optimizer. Of course, the PP/DS also has a much shorter planning horizon than SNP, so it is processing fewer planning buckets. Furthermore, PP/DS's scope is much narrower than SNP.

Conclusion

The objective of constraint-based planning is to generate a feasible supply and production plan. Production planning resources are not only used in PP/DS, but are also used in SNP, and are by far the most commonly-used resources in SNP. SNP and PP/DS share the same resources, which come over from SAP ERP. SNP and PP/DS use the same production resources, but plan differently, have different planning horizons and time orientations, and have different levels of detail to their planning. When used together, SNP and PP/DS must be synchronized to provide the desired output.

Furthermore, there are differences between SNP and PP/DS in terms of how much detail is included in the master data and planning runs, meaning that (and depending upon how much of this functionality is activated in PP/DS), SNP and PP/DS will not have the same planning output. Not much is written about this problem, but I will touch upon it in other chapters.

The hand-off between SNP and PP/DS is quite important, and the topic of planning horizons in both SNP and PP/DS were covered for this reason. In terms of timing, the trick is to configure the various horizons so that SNP and PP/DS are responsible for creating planned (production) orders for different times, although, as was explained in this chapter, there can be an overlap in this time between SNP and PP/DS.

CHAPTER 4

Constraint-based Methods in APO

My book *Supply Planning with MRP, DRP and APS Software* is dedicated to supply and production planning methods. In that book I combined coverage of the different supply planning methods with advice about—and historical analysis of—how to improve supply planning implementations. I also provided coverage of optimization in general, both outside and inside of APO. Because this material is covered in the book *Supply Planning in MRP, DRP and APS Software,* I will not repeat it here, but will cover only the methods that can work in a constrained manner in APO (optimization and CTM for both SNP and PP/DS), and will cover these in only a cursory manner.

The intent of SCM Focus books is to be tightly focused on one specialized topic, but also to have a number of books that cover supply chain planning from a wide variety of angles. This chapter provides a rather mechanical coverage of the settings that are available within the methods capable of constraint-based planning.

The CTM Profile

Capable to Match (CTM) is a form of heuristic-based allocation/prioritization supply and production planning method. CTM develops a supply

71

plan by allocating individual orders to planned inventory. Typically, allocation is based on some type of priority scheme (I say typically, because I know of several companies that use absolutely no prioritization scheme whatsoever with CTM).

Essentially, allocation assigns preferential treatment to high-priority customers at the expense of low-priority customers. High-priority customers can submit their requests for materials with the same requirement date as low-priority customers, weeks after the low-priority customers have submitted their requests. Even so, the high-priority customers will receive their allocation of stock. Many companies do not work on a first-come, first-served basis, and instead maintain extremely detailed management of customer priorities. In some industries, such as aerospace and defense, the military is a higher priority than civilian customers, and within the military customer category, different armed services have different priorities as well. However, the interesting thing about priorities is that the vast majority of companies do not use software to manage priorities, but instead use people.

Prioritization and allocation software such as CTM is an attempt to computerize these priorities. CTM works by creating a queue of customers, with the high-priority customers placed at the top of the queue and the low-priority customers placed at the bottom. Demands are then run through a supply-and-demand matching engine in the sequence specified by the prioritized customer list. Because the orders of the high-priority customers are run through before those of the low-priority customers, the planned inventory is allocated to the orders of the high-priority customers first. The orders belonging to low-priority customers are run toward the end of the sequence and there may or may not be inventory available for them when their demands are processed.

Let's get into the configuration details of CTM, many of which are held in the CTM Profile.

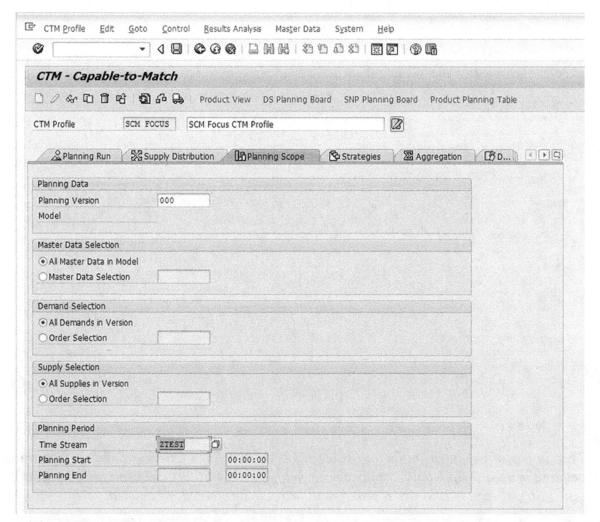

This tab is all about connecting the dots of master data, demands, supplies, and the time stream/planning calendar to be used.

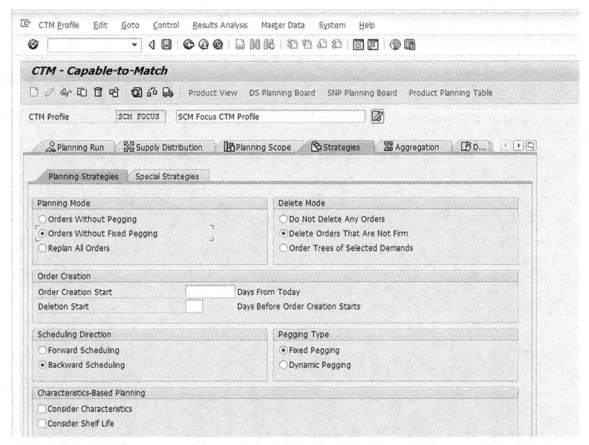

This is one of the most important tabs in CTM. It defines whether fixed or dynamic pegging is used (http://www.scmfocus.com/sapplanning/2009/05/06/pegging-in-scm/), scheduling direction and when orders will begin to be created, or deleted.

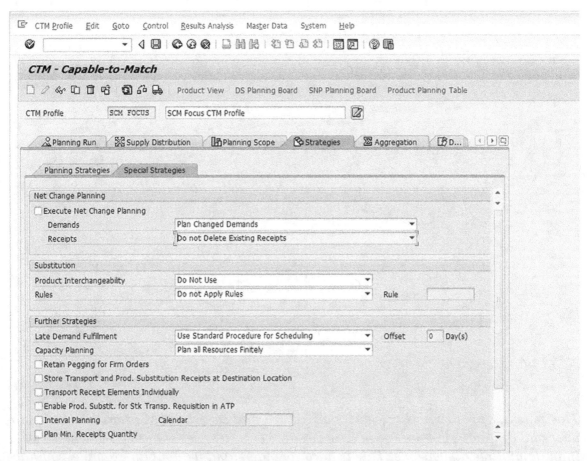

This tab contains a fine level of detail which controls things like whether planning should be net-change-based, whether interchangeability (SAP's term for product substitution) will be used, and whether peggings should be kept for firmed orders, as well as various other highly detailed settings that control the CTM run.

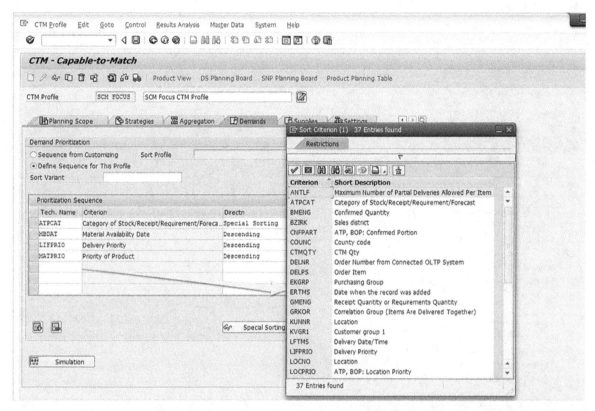

This is where the demand categories are added and sorted. CTM prioritizes both supplier and demands, but does so in terms of steps or as a series of nested sequences rather than in multiple dimensions. This is hard to understand without an example, so I have created one. For instance, in the prioritization sequence shown above, the sequence would follow the spreadsheet on the following page:

CTM Demand Sequence

This sheet is designed to simulate the sort sequence.

Objec Number	ATPCAT Sort Sequence 1	MBDAT Sort Sequence 2	LIFPRIO Sort Sequence 3	MATPRIO Sort Sequence 4
1	Sales Order	Material Availability 1	Delivery Priority 1	Product 1
2	Sales Order	Material Availability 1	Delivery Priority 2	Product 2
3	Sales Order	Material Availability 2	Delivery Priority 2	Product 2
4	Sales Order	Material Availability 2	Delivery Priority 4	Product 2
5	Forecast	Material Availability 1	Delivery Priority 2	Product 2
6	Forecast	Material Availability 4	Delivery Priority 2	Product 2
7	Forecast	Material Availability 4	Delivery Priority 2	Product 2

The first "criterion" is the order category. In this case, only forecasts and sales orders are being processed. Because I have defined a special sort profile where the sales orders are processed before the forecasts (shown on the following page), the sales orders are shown in the spreadsheet as coming before forecasts. However, before forecasts are processed, the orders by material availability dates are processed, and within material availability date, the delivery priority is sorted and processed, and within the delivery priority sort sequence the material priority is processed in sequence. Any sequence and sortation (even special sorting for non-numerical values) is possible for which a demand object exists. A list of all of the possible demand objects appears when the drop down for the Technical Name field is selected. The description—which has the heading "Criterion"—auto-populates after the technical name is selected. The only other thing left to do is to set the scheduling direction, which is ascending, descending, or special sorting. Special sorting is not a sortation direction, but instead is a pointer out to the special sorting configuration.

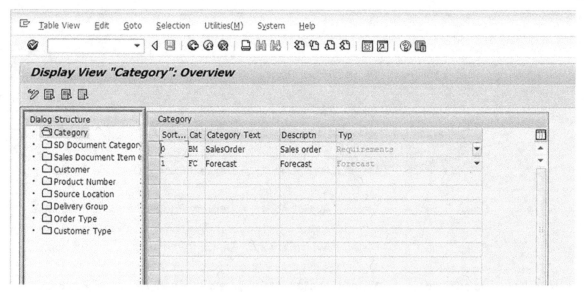

Here is the special sorting configuration screen. Any criterion can have a special sorting, which is necessary if the value is either not numerical or if ascending or descending will not meet the requirement.

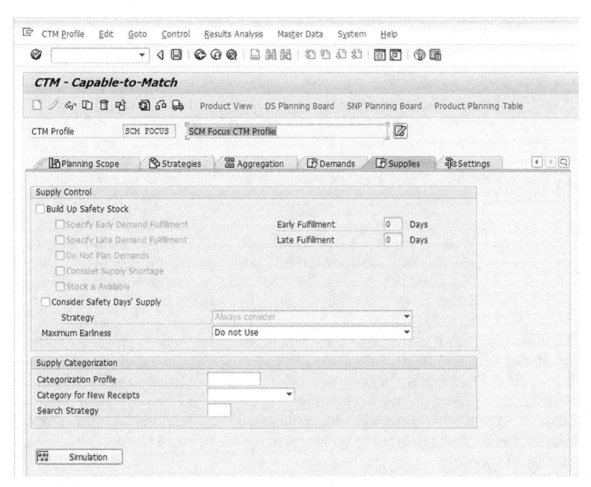

The Supplies tab has fewer options. The top portion of the screen deals with building up safety stock or considering safety days' supply, and how early or late the fulfillment should be allowed. While none of these fields have anything to do with supply prioritization, at the bottom of the screen there are three fields related to prioritization.

1. *Categorization Profile: This is where the supply category is assigned to the ATP category, and where the profile is set to either "use supply limits" or "ATP categories."*

2. *Category for New Receipts: With CTM supply categorization, you can assign a supply category to existing supplies and receipts. The system assigns the supply category specified here to new receipts, receipts that are created during the CTM*

planning run and are not completely consumed by the corresponding demand. This applies in the following cases:

　a.　You are using co-products.

　b.　You are using minimum lot sizes, fixed lot sizes, or rounding values.

If you do not specify a supply category here, the surplus receipts created during CTM planning receive standard category "00." CTM consumes receipts in this category before consuming receipts from other supply categories and before creating new orders. If you specify a supply category here that is not included in the search strategy used, the system cannot consume the surplus receipts created during CTM planning.

3.　*Search Strategy: Specifies the sequence in which CTM planning is to consume categorized receipts and supplies. The search strategy is maintained for a CTM planning run by assigning a sequence to each supply category. In order to fulfill a demand, the system then searches the supply categories in the sequence determined by the strategy.*

The main point of all of this is to allow you to prioritize some supplies over others. For instance, in the article link below, an example is provided of a configuration that would set the relative priority as stock-manufacture-purchase. This would be the normal sequence for most companies. However, if a company could purchase finished goods at a lower price than it could make them (making the main purpose of its own manufacturing make up for capacity-constrained times), the sequence could be stock-purchase-manufacture.

http://www.scmfocus.com/sapplanning/2012/06/22/order-or-supply-and-demand-selection-for-ctm/

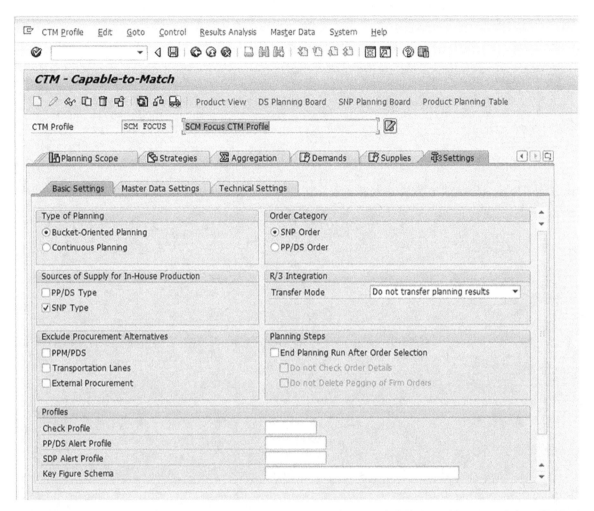

This tab contains the controls for the time orientation of CTM. When used for SNP, it will most often be "Bucket Oriented Planning." CTM can be used for PP/DS—that is, to create PP/DS orders directly so conversion is not necessary from SNP Planned Orders to PP/DS Planned Orders.

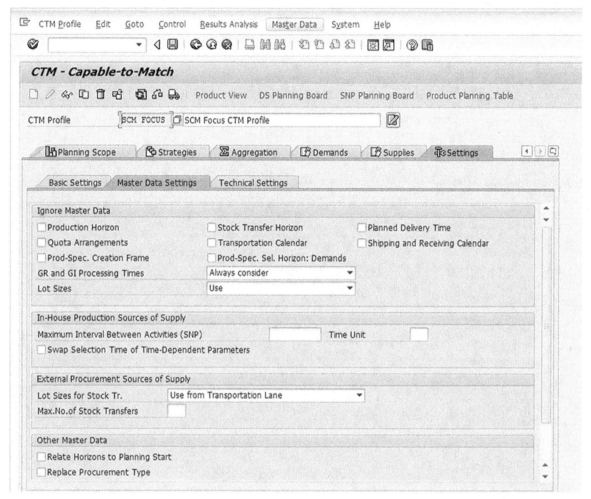

This tab allows different master data to be excluded from the CTM run. So for instance, if I wanted to ignore lot sizes, they would not be used by CTM but would still exist in the master data. I find this area useful from a simulation perspective. If a company wanted to know how the plan would be different if the quota arrangements that are set up in the system were not used, I can simulate that by making a copy of the active version to an inactive version, and then simply running the same CTM profile used for production, but with the quota arrangements excluded on this screen. (Or, as a setting in the next graphic will show, I can run this simulation right in the active version.) Through analyzing the planning output, and with some creative estimation, I can provide a company with a rough cost estimate for using quota arrangements.

The next set of controls relate to intervals between Planned Orders creation (which I don't use). The controls below this allow you to define where the lot sizes are pulled from. The options are Use from Transportation Lane, Use from Product—Destination Location, Use from Product—Source Location. So lot sizes could be defined in each place in the master data, and the setting simply switched per CTM run.

I do not use the rest of the settings on this tab.

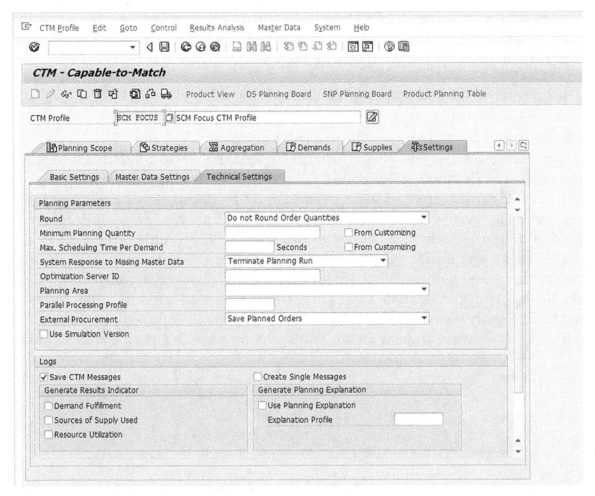

The final tab within technical settings controls the most detailed parts of the CTM run. CTM can be pointed to any of the Planning Areas in the System. This is also where the parallel processing profile is assigned. Parallel processing profiles allow CTM to take advantage of a multi-processor server by breaking the problem into threads. Parallel

processing profiles are almost always used in production because almost all servers are now multi-processor. Generally, the Optimization Server ID will be pointed out to the CTM01 Engine. The "Use Simulation Version" radio button allows the CTM planning results to be kept in a temporary buffer of the liveCache.

There are a number of other settings that make CTM tick, but I did not want to turn this into a book on CTM, as there is in fact already a book on CTM. Secondly, a number of the CTM settings have to do with prioritization rather than constraint-based planning, and prioritization is not the focus of this book. The link to all of the CTM articles at SCM Focus is below:

http://www.scmfocus.com/sapplanning/category/ctm/

The SNP and PP/DS Optimization Profiles

Optimization was explained in Chapter 2: "Understanding the Basics of Constraints in Supply and Production Planning," so instead of reviewing this information, we will jump right into the settings for both the SNP optimizer and PP/DS optimizers. SNP's cost optimizer can be used for both the creation of the initial supply plan as well as for the deployment planning run. The profiles that control the optimizer for the initial supply plan and for deployment are extremely similar, except that when running deployment there are no production constraints. Both the deployment and redeployment planning runs are covered in great detail in my book *Supply Chain Deployment and Redeployment in Software*. Therefore, I will only show the optimizer for the initial supply plan in the following graphic.

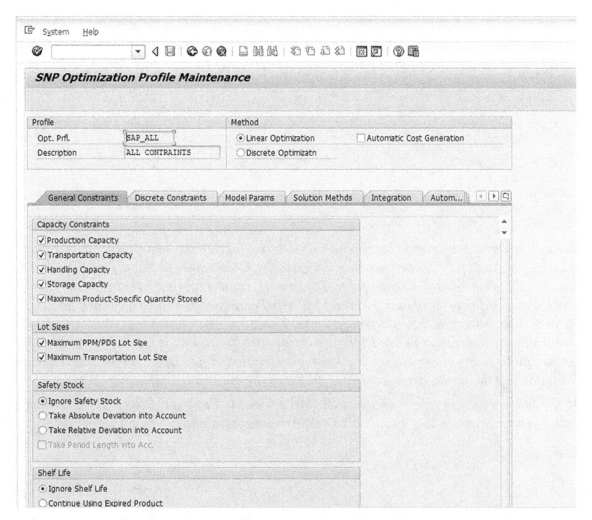

This tab primarily tells the optimizer what to include in its optimization run and what to exclude. I have not used the shelf life settings with the optimizer so I will not go over them.

The optimizer can be run to incorporate quota arrangements, which like prioritization (covered in Chapter 2: "Understanding the Basics of Constraints in Supply and Production Planning") is no longer a cost optimal approach (quota arrangements would then be selected even if the price of doing so were higher). Penalty costs can be assigned for exceeding or falling below the quota arrangements. These costs can be set low, medium or high, meaning that the company would have control over how much it wants the optimizer to choose quota arrangements over the lowest cost option. The higher the penalty costs for not following the quota arrangements, the more the optimizer will select them.

At the bottom of this tab, Supersession and Form-Fit-Function Class can be enabled. Both of these promote the optimizer to perform supersession.

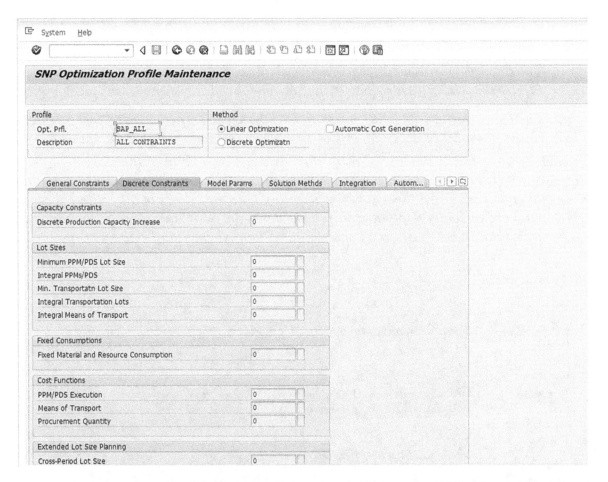

1. *Unlike continuous optimization, discrete variables are variables that are restricted in some way. For instance, when lot sizes are used in supply chain optimization, they "restrict" the results that can be computed. When the optimizer is asked to respect such batches (such as lot sizes), the optimization performs the solve with discrete variables. If no batching is used, the optimizer can select any quantity and the results that represent the solution are smooth, and are not stair-stepped. The discrete constraints are completely optional. However, the more discrete constraints that are included, the more realistic and usable the planning output will be. The rule with all of the discrete constraints is this: In order for the values that are placed in here to work, the Method on the header must be selected for Discrete Optimization.*

2. *The second section down is concerned with lot sizes. Lot sizes can be set in time units. So for example, the Minimum PPM/PDS Lot Size could be set to eight*

weeks, which is the horizon for when the discrete constraint—which is on the PPM/PDS—should be considered. After eight weeks it would not be considered.

3. *The third section down is concerned with the consumption that is placed into the PPM/PDS. This field also defines the horizon for which to consider the consumption.*

4. *The fourth section down is concerned with horizons for considering cost functions stored in the PPM/PDS: Means of Transport and the Procurement Quantity.*

In each of these cases, one can control the horizons for how long the discrete master data utilizes it. This makes for a more realistic plan in the nearer term, and a plan that runs more quickly toward the end of the planning horizon. Also, not all of these discrete constraints are typically populated in the master data, so if they are not populated in the master data, then the discrete constraints on this tab would also not be populated.

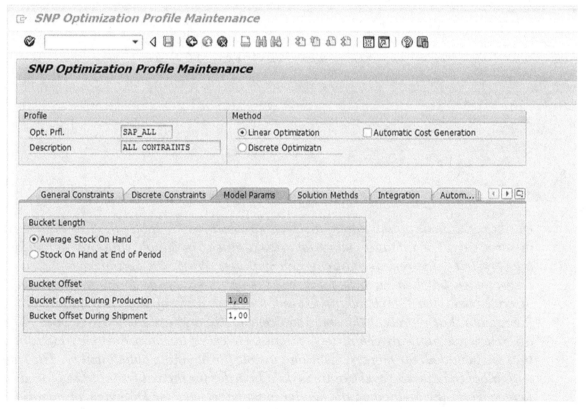

The Model Parameters tab controls just a few timings related to the bucket and how the stock is calculated, as well as the bucket offset that calculates the availability date of

the receipt element within a period. As stated by SAP, "the greater the value, the more optimistic the view because it implies that the products tend to be available earlier than specified."

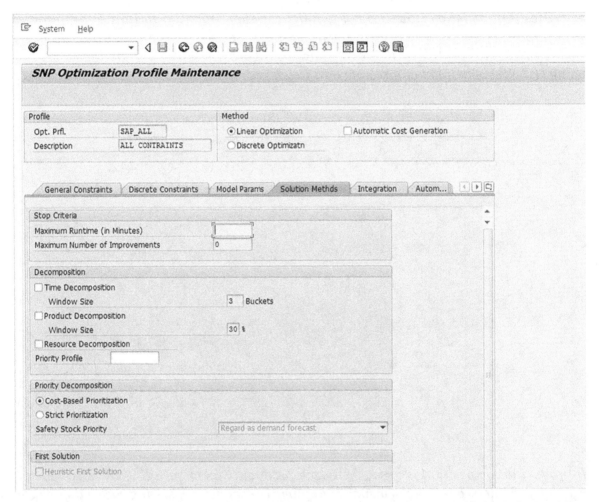

The Solution Methods tab shows how long the optimizer should run. Unless the problem to be solved is very simple, in most cases the run time of optimizers must be capped so that they can complete within the time windows available. The weekend provides a good chunk of time to run an optimizer, but most companies seem to want the results updated daily, hence requiring daily optimization runs—which of course reduces the available time in which to run the optimizer. If an optimizer were not capped it would eventually run until all the sub-problems were solved, which is the topic of the next set of controls on this tab.

The decomposition method is how the overall problem is divided into what are called sub-problems. There are three alternatives for decomposition: time, product, and resource. Product and resource decomposition cannot be used together, but time decomposition can be used with either product or resource decomposition. Decomposition can be used with priorities that go beyond the costs of non-delivery and allows priorities to be assigned to important products. The priorities for both products and resources are defined in the Define SNP Priority Profile. This is a simple configuration screen that assigns a priority value (1,2,3, etc.) to a list of products or a list of resources.

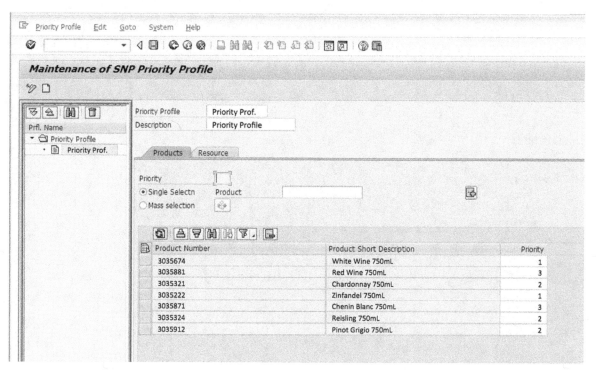

If product decomposition is used, then each product is taken for all of the locations for which it is valid as a sub-problem, and the optimizer goes to work solving each sub-problem. If resource decomposition is used, the problem is divided—with the resources becoming the sub-problem—and solved that way. Resource decomposition is preferable if the focus of the company is to load resources in a similar sequence. It also puts the resource at the center of the equation by solving the problem per resource. Resource decomposition can also be used with priorities. This imposes a specific priority; if no priority is used then the priority of the PPM/PDS is used by the system.

Further down on this screen the topic of priority comes up again. The standard way to run the optimizer is "Cost-Based Prioritization," where the penalty costs, which are set at the Product Master, determine the priority.

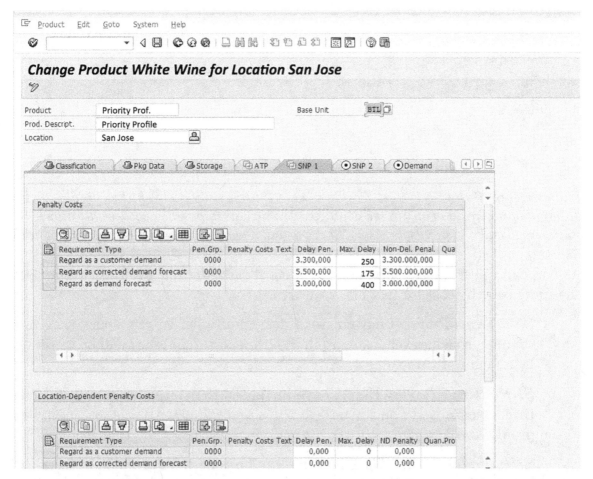

The SNP 1 tab holds both delay and non-delivery penalties. Because these penalties can be applied at the product level, some products can be prioritized over others based upon differentiated non-delivery penalty costs. As with the other costs added to the optimizer, these costs are traded off against other costs. These costs are not necessarily the final costs that the optimizer uses because SNP has something called cost profiles, which can be used to further adjust the cost values that are distributed in many different locations.

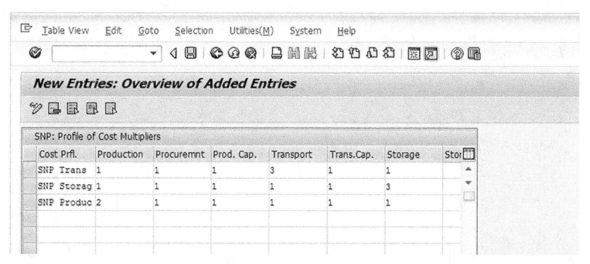

The existence of the SNP Cost Profile means that one must check this profile during different optimization runs to see what cost profile was applied. A cost profile allows the optimizer to quickly change its costs without actually touching any of the original cost data.

As you can see from the screenshot, I have three different profiles saved. Each one increases the cost for just one cost area.

- *SNP Trans Profile:* I have increased the multiplier for transportation costs to three, so whatever costs are in the transportation lanes will be increased by a factor of three.

- *SNP Storage Profile:* Here I have increased the storage cost multiplier by three, meaning the optimizer—for runs associated with this cost multiplier profile—will be run with the storage costs increased by that factor.

- *SNP Production Profile:* The same can be done with production costs as I have with this profile.

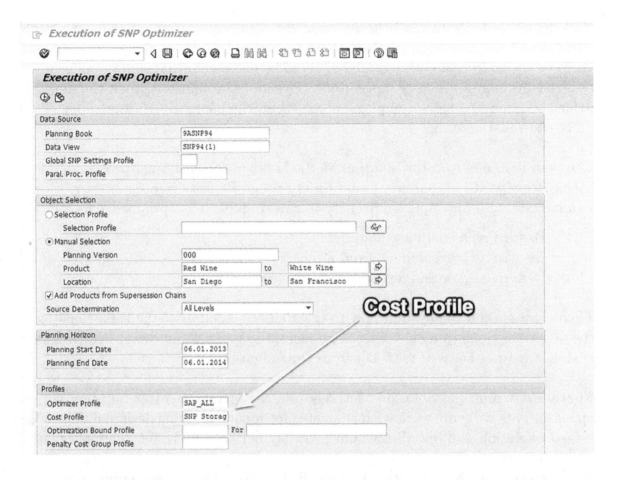

The cost profiles are assigned to the Execution of SNP transaction. The cost profile (as well as other profiles) that ran with a particular optimizer can be determined by looking into the optimizer log: /SAPAPO/SAPAPOLMSDP_OPTLGO.

However, the second option is to go with "Strict Prioritization," where the optimizer instead makes its decisions based upon priority of the demand (customer demand versus forecast, etc.). SNP has a series of default values for the priorities of each of the different categories of demand.

- "1" for customer demand
- "5" for corrected demand forecast
- "6" for forecast

This brings up an interesting question for companies that want to prioritize demands. That is, which method should be used? A company could use either CTM or the SNP cost optimizer. SNP could be used when the demand prioritization requirements are very basic, and CTM could be selected when the demand prioritizations are more complex and/or when the company wants to prioritize more than just demand.

The next item down on the Solution Methods tab of the optimizer profile is Safety Stock Priority. It is necessary to set Safety Stock Priority in order to run "Strict Prioritization." This applies a priority to safety stock. The options are as follows:

- Regard as a customer demand
- Regard as corrected demand forecast
- Regard as demand forecast

These options tell the system how to treat safety stock vis-à-vis the priorities of the demands. Something else that is important to know about Strict Prioritization is that it cannot be used with Discrete Optimization.

Further down on the Solution Methods tab, the Heuristic First Solution option appears. This will only work with Discrete Optimization and is designed to provide a faster solution, but one that would probably not be used in production.

Further down we have the LP Solution Procedure, with the following options:

- Primal Simplex Algorithm
- Dual Simplex Algorithm
- Interior Point Method

The first two options are the main methods available for the SNP Optimizer. Both of the options are "simplex," which is a form of linear programming first developed and published by George B. Dantzig. It was called the "simplex method" when first introduced and is now considered the "standard method" in optimization. "Dual simplex" is a modification of the simplex method that adjusts the problem and solves it differently.

The Interior Point Method is a modification of the two main methods. It uses an algorithm at the end of the solution process. These settings are not specific to SAP or even to the IBM/ILOG optimizer that resides within SNP, but generalizes the topic of linear programming/optimization. The Interior Point Method works differently than the standard simplex method because "contrary to the simplex method, it reaches an optimal solution by traversing the interior of the feasible region."[6] However, the mathematics of exactly what the Interior Point Method is doing is very much beyond the scope of this book, and I am certainly not the best person to explain this topic. Most implementers will want to test these settings to see if they improve either the solution or the runtime of the optimizer. Most of the optimization projects I have worked on have used the Primal Simplex Algorithm and never got to the point of testing these different options.

The next set of options relate to the topic of aggregation, which works with the resource hierarchies set up in the system. I discuss this in Chapter 5: "Resources." The options are as follows:

- Aggregated Planning—Vertical: Restricts the size of the model to be optimized. Here the product-location combinations are aggregated during the optimization and then disaggregated.
- Aggregated Planning—Horizontal: Used for planning certain product locations of the supply network.

For a full explanation of aggregated planning see my book *Aggregated Planning in Supply Chain Software*.

The final set of options relate to the objective of the optimization.

- Cost Minimization: This is the main way of running the optimizer. The objective function is to minimize costs, and the optimizer trades off all of the costs that have been entered into the model along with things like the SNP Cost Profile, which modifies these costs.
- Profit Maximization: Here the optimizer interprets the penalty for non-delivery costs that have been entered as lost profits. This is a very literal translation of costs, and I have never seen the SNP or PP/DS optimizer

---

[6] Wikipedia.

set up in this way. I have also never seen this setting used on any SNP implementation.

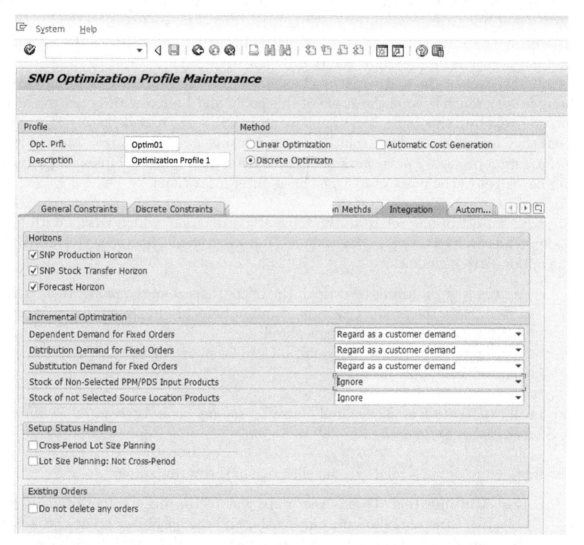

- SNP Production Horizon: The first three fields define whether the optimizer should use the horizons that have been placed into the Product Location Master. If any of these three options are not selected, then the horizon will not apply for the optimization run.

- SNP Stock Transfer Horizon: Time period in which SNP does not plan any stock transfers.
- Forecast Horizon: The horizon in calendar days during which the forecast is not considered as part of the total demand. (Details on this are related settings at this link: http://www.scmfocus.com/sapplanning/2011/06/09/forecast-consumption-setting/)
- Dependent Demand for Fixed Orders: These next three settings all relate to telling the optimizer how to interpret each of the different demands. The options, which assign different priority levels to each demand type are as follows:
  - Regard as a hard constraint: This is the highest priority—hard constraints must be met.
  - Regard as a pseudo-hard constraint: The next step down from a hard constraint.
  - Regard as customer demand
  - Regard as corrected demand forecast
  - Regard as demand forecast: This is the lowest priority—puts dependent demand at the level of a forecast.
- Distribution Demand for Fixed Orders
- Substitution Demand for Fixed Orders
- Stock of Non-Selected PPM/PDS Input Products: These next two settings can include stocks at source locations when planning. Applies for both production (PPMs/PDSs) and stock. Why would this be necessary? One reason could be that the source locations were left out of the master data selection when the optimizer was run. The options for both settings are the following:
  - Ignore: Does not activate this functionality.
  - Regard as a pseudo hard constraint: Activates this functionality.
- Stock of Non-Selected Source Location Products
- Do not delete any orders: This is the "net change" setting in the optimizer. If selected, the orders from the previous optimizer runs—even those that are not fixed—are not deleted. Fixing is ordinarily the method of ensuring that subsequent optimizer runs do not delete/overwrite the orders; however, with this setting the protection for fixed orders is extended to all orders. This is very basic net change functionality, which will not meet all of the

requirements for net change at companies. For instance, some companies only want to rerun the optimizer for locations that have changed since the last optimization run; a good example of this could be an updated final forecast. This setting will not do this. If this setting is activated, it will not delete an order and create a new larger order (in the event that the forecast was increased), but instead will keep the old order and then make a new order to meet the higher demand. This would be undesirable if the company preferred to have consolidated orders.

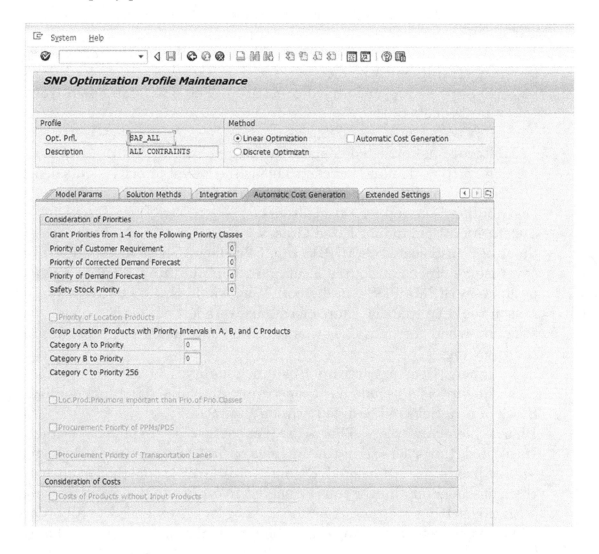

Most of the settings on this tab have to do with automatic cost generation. Automatic cost generation is a completely different and lower maintenance approach to optimization than using costs. When these settings are used, even if costs are entered into the model, the optimizer will not use them. Therefore, companies that have already populated costs can test automatic cost generation by simply populating the values on this tab. When these values are populated, the optimizer still uses its optimization objective function, but uses an interpretational layer to convert the priorities already entered.

- Prioritization of Customer Requirement: For the first four settings, priority is set as 1-4.
- Priority of Corrected Demand Forecast
- Priority of Demand Forecast
- Safety Stock Priority
- Priority of Location Products: These next three settings, which assign categories to priority values, can be used in addition to the first four settings. The first step to using the next three settings is to activate this setting, which instructs the optimizer to consider the priority of the location products. The priority of the location products is set in the Product Location Master. The priority setting causes the optimizer to move orders for products with higher priorities to earlier periods and vice versa. Values entered into the Product Location Master can be from 1 (the highest) to 255 (the lowest).
- Category A to Priority: For these settings, one must define the interval. For example, if 100 were entered into this field, the priority values higher than 100 would be considered Category A.
- Category B to Priority: If 150 were entered in this field (and assuming, as I have listed above, that 100 were entered into Category A to Priority), then Category B would be from 150 to 100.
- Category C would then automatically be from 100 down to 255.
- Location Product Priority more important than Priority of Priority Classes: As you can see, the previous settings have established two competing priorities. This setting instructs the model to give more emphasis to the second set of settings than the first. If this setting is not activated, then the first set of settings receives more emphasis.

Although, these next three settings are on the Automatic Cost Generation tab, they are not related to Automatic Cost Generation; they are related to priority.

- Procurement Priority of PPM/PDS: Causes the optimizer to consider priorities that are assigned to PPM/PDSs. Priorities must be assigned to PPMs/PDSs already in order for this setting to work. Priorities are the primary manner in which SNP and PP/DS make decisions between different PPMs/PDSs.
- Procurement Priority of Transportation Lanes: When this setting is activated, the optimizer considers the procurement priorities of the transportation lanes. This uses the setting in the Product-specific Transportation Lane. This setting is similar to the setting for PPMs/PDSs listed above. It allows the system to choose from several transportation lanes that have different priorities.
- (Storage) Costs of Products Without Input Products: Enables the optimizer to incorporate storage costs instead of the default value of "1" for the raw materials (that is, products without input products). Storage costs are set on the Procurement tab of the Product Location Master. By setting a high value, the system would postpone bringing raw materials into the supply network, or vice versa.

Because of the significant problems that companies have in determining their costs (and I have yet to see a company with a very good justification for the cost settings that they had set up), it is surprising to me that I have never seen the Automatic Cost Generation tab used. Furthermore, SAP does not really promote the use of automatic cost generation. If a company decides it would like to use the optimizer, to me it seems that using automatic cost generation would be a good place to start. In a prototype environment, the automatic cost generation could be run quickly for several different alternatives (with various changes to the six priorities on this tab) and of course with priorities set at the Product Location Master. The client could then review the output and decide if the automatic costs can meet their objectives. If not, they can move on to using costs and using the optimizer in its more standard implementation.

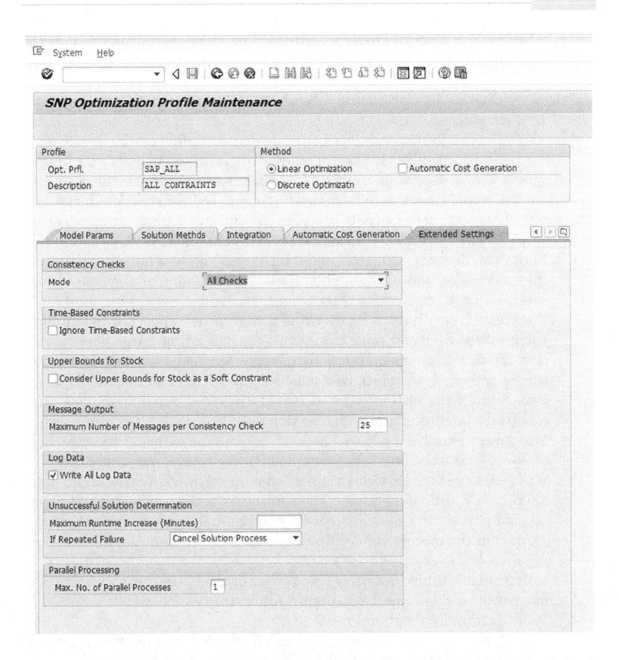

These are very detailed settings, and most of them are a bit esoteric. The default values will be used for most of the settings on this tab. However, I went through them because sometimes reviewing SAP settings is a bit like panning for gold: you end up going through a lot of sand with the potential to find some specs of gold.

- (Consistency Check) Mode: Determines how you would like to run the consistency check. The options are the following:
  - All Checks: All checks are made
  - Checks for Errors and Warnings: Checks are made to find error messages
  - Checks for Errors: Checks are made just to find errors
  - No Checks: No checks are made—turns off consistency checking
- Ignore Time-based Constraints: Used when you want to use the optimizer with a bound profile while there are time-based constraints in the system. This setting essentially prevents the optimizer from using two types of constraints, which are inconsistent with one another. A bound profile limits the decision variable deviations from the last optimization run, thus increasing planning stability. However, the bound profile is used very rarely on projects.
- Consider Upper Bounds for Stock as a Soft Constraint: You specify how you want the SNP optimizer to take into account a time-based upper bound that you may have set in interactive Supply Network Planning. If this setting is activated, the optimizer considers the stock upper bound as a soft constraint; if the setting is not activated, the stock upper bound is a hard constraint (meaning it may not be violated).
- Maximum Number of Messages per Consistency Check: Exactly as it says—restricts the number of messages to the consistency check.
- Write All Log Data: Performance can be improved by not writing all log data.
- Maximum Runtime Increases: The runtime is set on the Solution Methods tab, but this setting asks you to set a maximum increase in minutes in the event that the system cannot find a result for a whole or partial solution. I don't recall ever using this setting.
- If Repeated Failure: Specifies how you want the optimizer to proceed if it has not been able to find a solution. Options are as follows:
  - Terminate Solution Process.
  - Ignore Problem: Launches a search for a solution to the whole problem.
  - Simplify Problem: Ignores all discrete restrictions.
- Maximum Number of Parallel Processes: Determines how many problems the optimizer may create. This should be correlated to the free slots available for processing.

The PP/DS Optimization Profile

The PP/DS Optimization Profile is similar to the SNP Optimization Profile, but is even more detailed. These optimization profiles are one of the more complex areas of the PP/DS optimizer setup, and making changes to them requires a thorough understanding of how the PP/DS Optimizer works. Changes made to the standard profiles that ship with PP/DS can also require significant testing.

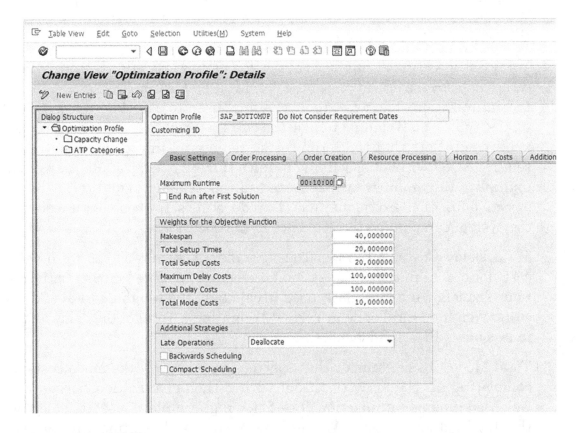

The following fields apply:

1. Maximum Runtime: This is how long we want the optimizer to run. Generally optimizers are capped at a specific time which fits within the overall operational workflow and available windows.

2. Makespan: A number that specifies how heavily the makespan is weighted during optimization. The weighted makespan is calculated as follows:

Weighted Makespan = Weighting x Makespan in Seconds. To reduce this time span, the system can change the sequence of the orders.

3. **Total Setup Times:** A number that specifies how heavily the total of the setup times for the operations in the optimization horizon is weighted during the optimization. The following formula is used to calculate the weighted total setup times: Weighted total setup times = Weighting x total setup times in seconds. To reduce the setup times for a resource, the system can change the sequence of the operations.

4. **Total Setup Costs:** A number that specifies how heavily the total of the setup costs for the operations in the optimization horizon is weighted during the optimization. The following formula is used to calculate the weighted total setup costs: Weighted total setup costs = Weighting x total setup costs x 1 second. Note that the setup costs are entered in the setup matrix as relative costs (without specifying a unit). During optimization, the system multiplies this number by one second so that the individual terms in the objective function are comparable. To reduce the setup costs for a resource, the system can change the sequence of the operations.

5. **Total Delay Costs:** A number that specifies how heavily the total delay costs for orders in the optimization horizon are weighted during optimization. The following formula is used to calculate the weighted total of delay costs: Weighted total of delay costs = Weighting x Total of the delay length in seconds x Priority-dependent costs.

6. **Total Mode Costs:** Number that specifies how heavily the mode costs for completing all the activities from within the optimization horizon are weighted during optimization. The following formula is used to calculate the weighted mode costs: Weighted mode costs = Weighting x mode costs in seconds.

7. Late Operations: Determines how optimization handles operations that are late for the requirement date/time. The following options are available:
   a. Schedule late operations: If this option is selected, the optimization schedules all operations.
   b. Deallocate late operations, if they end after the end of the optimization horizon: If this option is chosen, the optimization deallocates operations that end after the end of the optimization horizon, and positions them as late as possible within the optimization horizon.
   c. Deallocate all late operations: If this option is selected, the optimization deallocates all operations that are late for the requirement date/time or end after the end of the horizon.
   d. Deallocated Orders: If the Only Scheduled option is selected for the Scheduling Status of Operations field on the Order Processing tab page in the optimization profile, orders deallocated before the optimization-run retain their planning status.

8. Backwards Scheduling: Specifies that the system executes backward scheduling at the end of the actual optimization. By doing this, the system tries to schedule the activities that were scheduled too early, as close as possible to the requirement dates.

9. Compact Scheduling: Determines that the system executes compact scheduling during optimization. In doing so, the system tries to reduce the lead time of the orders—that is, to plan this per order with the smallest possible amount of time between the activities. The objective function criteria Setup or Lateness are not made worse by this. With compact scheduling, additional orders are not delayed.

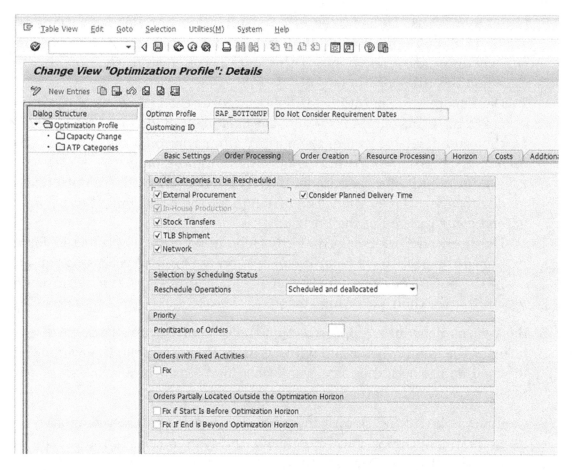

1. **External Procurement Orders Can Be Rescheduled During Optimization:** If you set this indicator, the system can reschedule external procurement orders (purchase requisitions) during optimization.

2. **In-House Production Orders Can Be Rescheduled During Optimization:** If you set this indicator, optimization can reschedule in-house production orders. This indicator is set automatically if the external procurement indicator was set.

3. **Stock Transfers Can Be Rescheduled During Optimization:** If you set this indicator, the system can reschedule stock transfer orders during optimization.

4. TLB Shipments Can Be Rescheduled During Optimization: If you set this indicator, the system can reschedule transportation orders during optimization.

5. Network: If you set this indicator, the system can reschedule project orders during optimization.

6. Consider Planned Delivery Time / In-House Production Time: With this indicator you determine that, during optimization, the system is to consider the Planned Delivery Time or the In-House Production time for all products that are procured externally or produced in-house (components/assemblies) and that have shortages.

7. Scheduling Status of Operations: Specifies which operations are to be selected based on their scheduling status for the optimization run. Choose a scheduling status.
   a. If you **also** select Deallocated Operations for Optimization, these operations then receive the status scheduled by the optimization run.
   b. If you select **only** Deallocated Operations, the scheduled operations are not changed by optimization.
   c. If you select **only** Scheduled Operations for Optimization, the deallocated operations remain deallocated, but are rescheduled so as not to restrict the optimization of scheduled operations by dependencies.

8. Preference of Determined Order Priorities: Using Order Priority, you can set which orders should be scheduled with greater preference. The system divides the available orders into two groups: the group with orders that meet the given order priority, and the group with orders that do not meet the given order priority. The orders that meet the order priority are rated as very important and are scheduled with greater preference during optimization.

9. Fix Orders with Fixed Activities: If you set this indicator, orders that have at least one fixed activity are fixed during optimization. The following objects are fixed in principle during optimization:
   a. Manually fixed, started, partially confirmed, or completely confirmed operations.
   b. Operations that do not rely on the resources selected for optimization.
   c. Activities outside or only partially within the optimization range.

10. Fix Orders that Start Before the Optimization Horizon: If you set this indicator, those orders that begin before the optimization horizon and extend into it are fixed during optimization. Therefore the activities of these operations cannot be rescheduled during optimization.

11. Fix Orders that End Beyond the Optimization Horizon: If you set this indicator, those orders that begin in the optimization horizon and last after it has ended are fixed during optimization. The activities of these operations cannot be rescheduled during optimization.

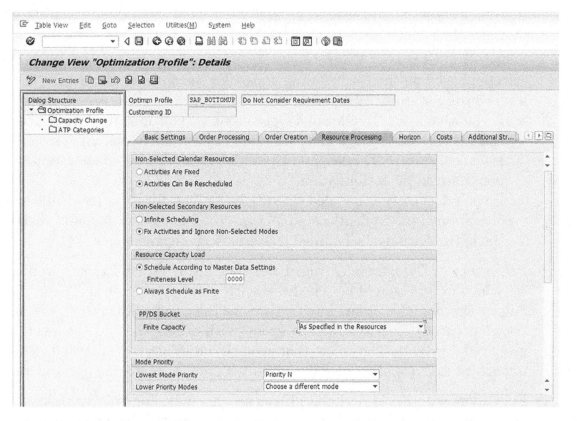

1. Activities on Nonselected Calendar Resources are Fixed: Specifies that in-house production order activities on calendar resources that were not selected for the optimization-run are fixed; that is, they can no longer be moved by optimization. The system uses this setting by default.

2. Schedule Nonselected Secondary Resources Infinite: Determines that optimization also plans nonselected resources as infinite, if they are used in an activity as a secondary resource. Nonselected resources do not result in restricted mode selection for optimization. Use this setting if you do not have to plan secondary resources as finite or if finite planning for these resources is not necessary until a later planning step. In this way you can ensure that optimization is not too restricted and can attain better results.

3. Schedule Resources as Finite or Infinite: Specifies that resources are to be planned as finite or infinite, as in the Resource Master. This setting is used by the system as a standard procedure. If you use this setting, optimization schedules activities to resources according to how you defined them in the resource planning parameters.

4. Always Schedule Resources as Finite: Indicates that resources always have to be scheduled as finite. If you use this setting, optimization always dispatches activities to resources with a finite capacity load, even if you have not set the Finite Scheduling indicator in the planning parameters of the Resource Master.

5. Finite Capacity Resources with PP/DS Bucket: Determines which capacity of the resource is to be used for finite scheduling in PP/DS.

6. This field is only relevant for bucket-oriented capacity planning in the CTP process and in bucket-oriented block planning. It only affects the single-activity and multi-activity resources and the single and multi-mixed resources that have PP/DS bucket capacity.

7. Lowest Mode Priority: This is the lowest mode priority that the system should consider during automatic mode selection. For example, if you specify the lowest mode priority C, the system only considers modes with priorities A, B, and C; meaning that it only schedules operations on the resources belonging to these modes. If you do not specify a value for the lowest mode priority, the system considers all modes that can be scheduled by the system (modes with priorities A to O).

8. Lower Priority Modes: In this field, you specify how you want the system to handle operations during optimization that are scheduled on a mode with a lower priority than that specified in the Lowest Mode Priority field.

9. Consider Resource Buffer for Pegging: You can choose whether you want optimization to take into account the time buffer of the respective primary resource when scheduling activities that have pegging relationships between them.

10. Schedule Relationships Use: You can specify that you want relationships based on routing data from SAP R/3 to be scheduled in SAP APO. You have the following options:
    a. Optimization is to schedule these relationships: The system is to include availability of scheduled resources when it calculates the minimum and maximum intervals between the activities connected by these relationships.
    b. Optimization does not schedule these relationships.
    c. Optimization only schedules relationships on resources that are planned on an infinite basis.

11. Consider Resource Networks in the Optimizer: You can choose how the optimization should consider the resource network connections:
    a. Ignore: Resource networks will not be considered.
    b. Order internal only: Consider resource network connections, and only order internally.
    c. Cross order only: Consider resource network connections only between orders.
    d. Always: Consider resource network connections always (order internally and between orders).

12. Lowest Resource Network Priority: Defines the lowest priority of the resource network connections the optimization can use. If you do not specify a value for the lowest resource network priority, the system considers all resource network connections that can be scheduled by the system (priorities A to O). If you specify the lowest resource network priority C, the system only considers resource network connections with priorities A, B, and C.

13. Considering Storage Properties During Optimization: If you set this indicator, the system activates container planning. Optimization then considers container resources regarding the maximum fill level and the settings for product purity according to the Resource Master data.

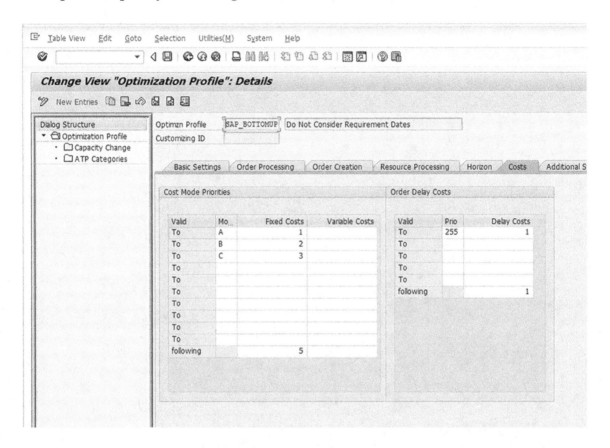

Conclusion

This chapter covered the configuration in SNP and PP/DS of the methods that can perform constraint-based planning. Technically, these available methods are heuristics, CTM, and optimization; but in practice, CTM tends to only be used for SNP and not for PP/DS. CTM and optimization have a tremendous number of settings, so a good percentage of this chapter simply explained how these different settings work.

CHAPTER 5

Resources

Different domains of the supply chain have different types of resources. For instance, trucks are a resource for supply planning, while a workstation is a resource for production planning. However, all resources work similarly in that they have a capacity that can be declared to the planning system. With unconstrained planning (or infinite planning), capacities may or may not be declared, and, if they are declared, there is nothing to stop the system from placing an unlimited load on any resource. With constraint-based planning, three things happen: (1) resources are both declared; (2) at least one of the resources in each process chain has its capacity constrained or capped; (3) the system can only load the resource up to that cap before moving further loads to a different time or to an alternative resource.

Of the different supply and production planning methods available in APO, only prioritization/allocation (CTM) and cost optimization have the ability to run in a constrained fashion. However, this occurs only if the system is configured to manage resources in a finite manner. It is important to understand the distinction between what any system is capable of doing and what it is actually doing in a specific implementation. (This mistake is made in many areas of software evaluation. For instance, it is often stated that an application has a capability of doing something, but this does not say anything about how effective the functionality is or how difficult or easy it is to configure or to maintain.)

The Resource Types

Resources are the mechanism for both constraining the plan in SAP APO and determining if a plan is feasible. Resources apply to SNP, PP/DS, and TP/VS. However, a resource type is the category of the resource, and because different resources do different things, there are different resource types in SAP APO. The most common resources used in SAP APO are production resources, and this applies both to SNP and PP/DS. Therefore, SNP creates the "initial production plan," in addition to creating the initial or network supply plan. Then PP/DS, when deployed (not all companies that implement SNP implement PP/DS, and not all companies that implement SNP implement a production planning and scheduling system—but most do), follows the supply plan and produces a more detailed production plan and finally a production schedule. SNP can plan down to the daily bucket; however, within the day, the supply planning system lacks visibility. When planning occurs for production within the day, this is called scheduling, and is only performed by the DS portion of PP/DS. There are four types of resources in APO that apply to SNP and PP/DS, or to supply planning and production planning. A link to an article is provided below for those interested in more on this topic.

http://www.scmfocus.com/sapplanning/2009/05/02/scm-resource-types/

The following resources are available within APO:

1. *Storage:* Used to model the storage capacity within locations. Storage resources are occasionally used, and I have been asked to activate them on projects. I have also seen several projects where storage resources were capacity-constrained, but in both cases the constraint was deactivated after a brief period because in reality the companies did not want to actually constrain based upon storage capacity. Storage resources cannot be capacity-leveled. Therefore, in the several cases I have seen storage resources used, they have primarily been used for visibility.

2. *Handling (Unit):* A resource is intimately tied to the goods issue (GI) and goods receipt (GR) process. When it is used, it is more as a necessary configuration item to achieve another objective such as using the GR/GI processing time in a particular way. In reality, handling resources are inexpensive and generally not considered strategic. While there are exceptions for specialty products, handling resources consist primarily of things like warehouse workers, forklifts, roller conveyers, etc. More on this is explained at this article link:

   http://www.scmfocus.com/sapplanning/2012/12/06/the-goods-receipt-processing-time-and-the-handling-resource/

   However, I will later describe how a handling resource can be used to model the overall processing capacity of a location for both inbound and outbound processing.

3. *Transportation:* Transportation resources can be used by either SNP or TP/VS. I don't cover TP/VS in this book and TP/VS has very few implementations globally, even though it is one of the original five APO modules. Furthermore, I have never seen transportation resources used in SNP and there is a good reason for this, which I will describe in more detail later in this chapter.

4. *Production:* This option models production capacity in the factories. These are by far the most used resources in APO. They are not only used in PP/DS, but are also the most commonly used resources in SNP. In fact, I cannot recall an SNP implementation that I either implemented or saw after the fact that did not use production resources.

So now that we have reviewed the different resource types (every resource that is created in APO must be assigned to one of these resource types), we can discuss what these resources represent. Some specific examples of supply chain planning constraints include the following:

- The maximum amount of material that can be placed in a truck
- The maximum amount a bottleneck resource on a production line can produce in an hour
- The maximum number of miles a truck can drive in a day
- The maximum output available from a piece of machinery

- The maximum amount a bin at a warehouse can hold
- The maximum amount of material in total a warehouse can hold
- The time windows (i.e., calendar) that is acceptable for material delivery for production facilities—warehouses and retail locations

The different resource types are apparent in both the resource setup, as well as the Selection Profiles in the planning book when resources are selected.

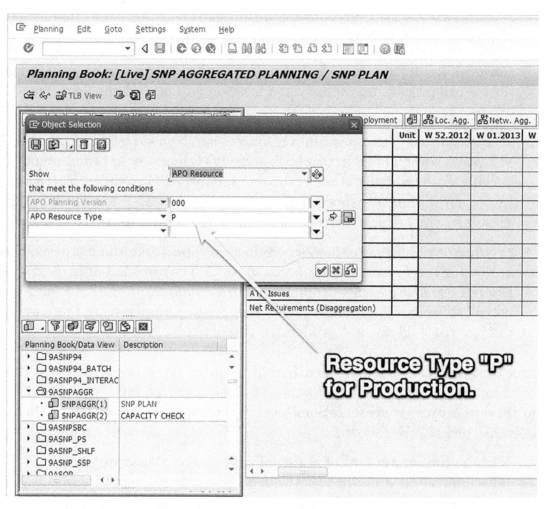

Here in the Selection Profile of the planning book I have defined a resource selection, and then asked to see only "P" for production resources. I can bring up all production resources by doing this.

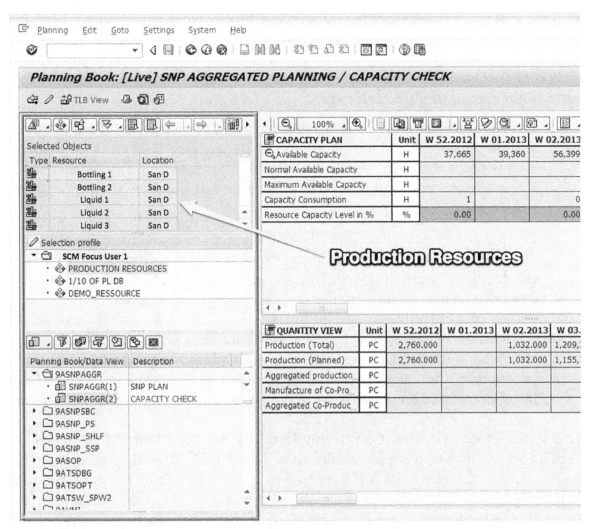

If I select all the production resources and then select the open folder, I can see all scheduled production on all resources. This can also serve as a report, as I can see the total resource utilization in this view.

Production Batch Sizes versus Full Resource Constraints

Supply planning software can represent production planning constraints by using lot sizes on internally-produced products. Doing so would serve to batch production orders for the supply plan when using any of the supply planning methods. A supply planning system could have lot sizing without incorporating any resources into the system. Furthermore, resource constraints, which are of course far more detailed than lot sizes, can also be included when using either the allocation or cost optimization methods. There are two main reasons why production constraints should be included with resource constraints in a supply planning system rather than relying upon production lot sizes:

1. *To Constrain the Plan to a Feasible Schedule:* Production lot sizes simply declare the batch in which an order must be produced (500 units, 1000 units, etc.) without declaring whether or not that lot size quantity is feasible, or if there is any capacity to produce that item requested by the supply planning system.

2. *To Model the Constraint:* This allows the ability to model the constraint and attach costs to the constraint in the event that cost optimization is used as the supply planning method. In cost optimization, both fixed and variable costs can be assigned to production. Costs must be assigned to any activity with a cost optimizer if the intent is to have the cost optimizer trade off the activity in question with other activities that are the supply planning recommendations. Although not all supply planning methods incorporate production and resource constraints, all supply planning methods can incorporate production realities (if not their constraints) by using lot sizing.

The Types of Constraints Commonly Used by Supply Planning Systems

While constraint-based supply planning is discussed frequently, and many software vendors talk about the full complement of resources that they offer, the talk is usually about what software is capable of hypothetically, rather than how it is typically implemented. The various books or vendor manuals on resources would lead one to believe that a wider variety of resources are typically used in supply and production planning than actually are used. Therefore, the most common resources used in supply planning and why they are used are discussed much less frequently.

Constraint-based planning, an approach to planning that has been demonstrated to work well with respect to production resources, does not translate the same way to supply resources. Constraint-based planning actually began in production planning and scheduling, and then migrated into supply planning. However, supply planning and production planning are significantly different. Production planning models a sequence of activities that are implemented by resources. In discrete manufacturing, the activities are usually sequential (one resource completes a process and the next resource begins work on the material). In true process manufacturing (as opposed to mixing operations), the resources are still sequential, but they are lagged and the overall sequence is a good deal more complex than in discrete manufacturing. However, as long as a series of operations is sequential, production planning can be easily throttled or constrained by the resource with the lowest capacity along the line—the so-called bottleneck resource. The sequential and closely interdependent nature of the production line allows it to be effectively constrained by one resource, meaning that it is only necessary to model one resource along the entire line. I learned this rule through my project work, as books tend to describe the theory behind constraint-based supply planning rather than its practical implementation.

How Supply Planning Resources Differ from Production Planning Resources

Unlike production resources, supply planning resources (storage, location, shipping and receiving, and transportation) are not in a direct sequence. When each is used in reality, supply planning resources have substantial lags between them. For instance, stock will often stay in a warehouse for some time before being moved onto a truck and on to the next destination in the supply network. Factories have work-in-process between work stations/resources, and the lag is much less substantial. Rather than being part of a manufacturing process, supply planning is made up of distinct processes. For instance, goods issue, the transportation of products, the goods receipt, and put-away are really separate processes.

Clearly, the requirement of constraining production planning is much easier to meet than the requirement of constraining supply planning resources. Production constraints are also much more prone to being constrained and in fact are more important for a company to constrain than supply planning resources.

Therefore, the most common resources to be constrained in supply planning systems are not supply planning resources, but production resources. Both supply planning and production planning use production resources. SNP creates an initial production plan, which is at a higher level but which looks out considerably further than the PP/DS final production plan and detailed schedule. There are other differences. For instance, SNP will tend to use bucketed resources, while PP/DS will use time continuous resources (configuring mixed resources allows a single resource to be used by both SNP and PP/DS as it has both bucket and time continuous orientation tabs).

APO Resources in Detail

Resources represent the various capacities within both supply and production planning. SNP can use all the resource types (as SNP can use production resources in addition to supply planning resources), while PP/DS can only use production resources. Therefore, instead of covering resources separately in SNP and PP/DS, it made more sense—to me at least—to dedicate a chapter to resources.

Time Settings on the Resource

The following time-related settings can be found for the various resources that can be set up in APO.

- Single

- Single-Mixed

- Multi-Mixed

- Production Line

- Line Mixed* (Only used in IPPE, not frequently used)

- Bucket

- Vehicle* (Only used for TP/VS, so not covered in this book)

- Transportation* (Used in SNP, but infrequently used as transportation is rarely a hard constraint)

- Calendar* (Not frequently used)

- Calendar Mixed* (Not frequently used)

In this chapter I will cover timing-related fields for the resource types just listed, except those with an asterisk next to them.

While each resource has its own set of tabs and time-related fields, most of these fields are identical or similar. Rather than repetitively listing all the timing-related fields for each resource, I will list the similar timing fields for all of the resources, but note where there are differences. Additionally, I will highlight those resources that have timing fields that other resources do not.

First, let's define some of the resources that appear in this chapter and that are integral to understanding the different resources classifications. We will define the following types of resources:

- Bucket versus Time Continuous Resources

- Mixed Resources

- Single versus Multi Resources

- Single-Mixed versus Multi-Mixed Resources

Bucket versus Time Continuous Resources
A bucket resource has a quantity capacity that is planned on a daily basis and would be defined as something like "400 bottles per day." A bucket-oriented resource is based upon a Bucket Oriented Dimension, the options of which are listed here:

- Area

- Density

- Elec. Current

- Energy

- Force

- Frequency

- Length

- Mass

- Mass flow

- Power

- Pressure

- Speed

- Temperature

- Time

- Volume

After you have selected a Bucket Oriented Dimension, you then select the unit of measure for the dimension. This is different than a time-continuous resource, as noted below:

1. *Time Continuous Planning:* This is applied to PP/DS PPMs and PDSs and provides the highest detail available within APO. However, PP/DS can be set up as a bucket capacity, and in this case, it is necessary to set up a Single Mixed resource. The PP/DS Bucket Capacity tab exists on only this time-continuous planning resource type.

2. *Bucket Oriented Planning:* Bucket Oriented Planning uses the SNP PPM or PDS, which is more abstract.

Technically speaking, SNP can use time-continuous resources, in that they can be set up in the system. However, as SNP does not schedule below the day, it makes little sense to have SNP use a time-continuous resource.

This topic is explained further at the following link.

http://www.scmfocus.com/sapplanning/2012/08/19/time-continuous-planning-versus-bucket-in-ctm-and-ppds/

Mixed Resources

The nature of the bucket versus time-continuous resource also relates to the topic of the mixed resource. As I have stated, SNP uses bucketed resources, while PP/DS uses time-continuous resources. However, a mixed resource can be set with both bucketed time master data parameters and time-continuous master data parameters. SAP provides the following, highly illuminating quotation on the topic of mixed resources:

> *If you want to consider the resource loads caused by PP/DS orders in SNP planning, and adjust the SNP planning accordingly, you must use mixed resources (single-mixed resources or multi-mixed resources). In mixed resources, you define the bucket capacity for period-oriented planning in SNP and the time-continuous capacity for time-continuous planning in PP/DS. An SNP order utilizes the bucket capacity of a mixed resource and a PP/DS order utilizes the time-continuous capacity of a mixed resource. For SNP planning, the amount of bucket capacity utilized by PP/DS orders is displayed as an aggregated capacity requirement. SNP planning can therefore take account of the PP/DS orders. For PP/DS planning, the time-continuous capacity used by SNP orders is not displayed. — SAP Help*

Therefore, a mixed resource has both bucket and time-continuous master data parameters (both for timing—which are discussed here, and also for other settings not covered in this book). A mixed resource can be used for both SNP and PP/DS—that is two different applications, but one resource. However, each application uses the resource—or looks at the resource—in its own way, which helps to explain how production resources that are shared by SNP and PP/DS can be the same. This is a very common question on projects.

Single (Activity) versus Multi (Activity) Resources

After the difference between multi resources and non--multi resources is made clear, there is a distinction between single or mixed resource. A single resource only allows one activity to be performed at one time on the resource, while a multi-resource allows more than one activity to be performed at one time.

Single-mixed or Multi-mixed Resources

Single-mixed and multi-mixed resources work in the following way:

1. *Single-mixed Resource:* This resource can only process one activity at a time, but can be used by both SNP and PP/DS because this resource type has both time-continuous and time-bucketed master data. (Actually, this resource type can be used for time-bucketed planning for both PP/DS and for SNP, or time-continuous planning for PP/DS and timed-bucketed planning for SNP.)

2. *Multi-mixed Resource:* This resource can process multiple activities at a time, but can be used by both SNP and PP/DS because this resource type has both time-continuous and time-bucketed master data. (However, unlike the single mixed resource, this resource can only be used for time-continuous planning for PP/DS.)

Resource Settings and What This Means for Time-Continuous versus Bucket Planning

Different resource types, such as single-mixed and multi-mixed, can be set up to be used by both SNP and PP/DS. However, how the resources are used can change in SNP depending upon the setting. As is highlighted in the quote below, both time-continuous and bucket-oriented planning can use mixed resources.

> *If you want to consider the resource loads caused by PP/DS orders in SNP planning, and adjust the SNP planning accordingly, you must use mixed resources (single-mixed resources or multi-mixed resources). In mixed resources, you define the bucket capacity for period-oriented planning in SNP and the time-continuous capacity for time-continuous planning in PP/DS. An SNP order utilizes the bucket capacity of a mixed resource and a PP/DS order utilizes the time-continuous capacity of a mixed resource. For SNP planning, the amount of bucket capacity utilized by PP/DS orders is displayed as an aggregated capacity requirement. SNP planning can therefore take account of the PP/DS orders. For PP/DS planning, the time-continuous capacity used by SNP orders is not displayed. –**SNP Help***

The system can automatically derive the bucket capacity of a mixed resource from the time-continuous capacity. Since you do not plan with so much detail in SNP (for example, you do not use sequence-dependent setup times), you can reduce the bucket capacity derived using a loss factor. You obtain such a buffer for detailed planning in PP/DS.

For single-activity resources, multi-activity resources, and calendar resources, or for the available time-continuous capacity resources of mixed resources, always enter a rate of resource utilization of 100 percent and a break duration of 00:00:00. Otherwise liveCache determines a different duration than does CTM planning for the corresponding activities. This may cause the system to fulfill the demand too late.

> *CTM always prefers to plan activities overlapping on multi-activity resources and does not support the synchronization of activities. Ensure that the No Synchronization setting in the Resource Master on the Planning Parameters tab page under SyncStart is selected. Otherwise the SAP liveCache executes a synchronization for the corresponding activities. — **SAP Help***

This is shown in the graphic below, which emphasizes and reinforces these points.

Resource Types and Their Function

		Single	Single Mixed	Multi	Multi Mixed
Function	Can Represent Both Time Bucketed and Time Continuous Resources	No	Yes	No	Yes
	Can Perform More than one Operation at a Time	No	No	Yes	Yes
	Can Represent Both Time Bucketed and Time Continuous Resources, as well as Specifically Time Bucketed Resource for both SNP and PP/DS	No	Yes	No	No

Now that we have covered the basic commonly used resource types in APO, we can get into the next layer of detail, which is the timing fields that are on each resource.

Timing Fields Per APO Resource Type

		On Which Tab?	Single	Single Mixed	Multi	Multi Mixed	Production Line	Bucket
Fields	Time Zone	General Data	Yes	Yes	Yes	Yes	Yes	Yes
	Factory Calendar	General Data	Yes	Yes	Yes	Yes	Yes	Yes
	Days +	General Data	Yes	Yes	Yes	Yes	Yes	Yes
	Days 1	General Data	Yes	Yes	Yes	Yes	Yes	Yes
	Start	Time Contin. Capacity	Yes	Yes	Yes	Yes	Yes	No
	End	Time Contin. Capacity	Yes	Yes	Yes	Yes	Yes	No
	Break Duration	Time Contin. Capacity	Yes	Yes	Yes	Yes	Yes	No
	Synchronization Start	Time Contin. Capacity	No	No	Yes	Yes	No	No
	Time Buffer	Time Contin. Capacity	Yes	Yes	Yes	Yes	No	No
	Productive Time in Hours	Time Contin. Capacity	Yes	Yes	Yes	Yes	Yes	No
	{Base Rate) Per	Time Contin. Capacity	No	No	No	No	Yes	No
	Period Type	SNP Bucket Capacity	No	No	No	No	No	Yes
	Number of Periods	SNP Bucket Capacity	No	No	No	No	No	Yes
	{Valid From} Start	Downtimes	Yes	Yes	Yes	Yes	Yes	Yes
	{Valid To} End	Downtimes	Yes	Yes	Yes	Yes	Yes	Yes

As can be seen from the previous matrix, most of the resource types have the same timing fields. The exception here is the bucket resource, which has two timing fields. No other resource type has any of the Time Continuous Capacity fields.

Interestingly, the resources are given the option of either using a set-up matrix or a synchronized start.

The Time Orientation Flexibility of Capable to Match (CTM)

A good example of what we just covered is found in CTM. Within APO there are three methods for supply planning: the SNP Optimizer, the SNP Heuristic, and Capable to Match (CTM). However, of the three, CTM can also be configured to perform not only supply planning, but also production planning—creating planned orders that are native to PP/DS (rather than requiring conversion from SNP to PP/DS). CTM has this ability primarily because it has been designed by SAP to work in either the bucketed or the time-continuous orientation.

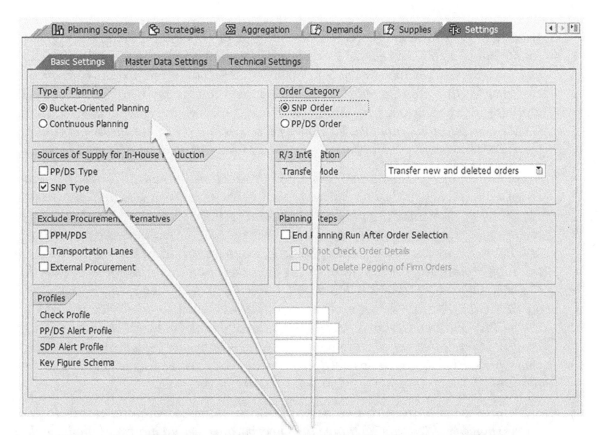

Here you can see that the CTM Profile can be set to run in either SNP "mode" or PP/ DS "mode." In order to switch between modes, the settings above would be changed. In the screen shot I have SNP Type planning enabled, but by switching to the other option

for each setting, I could convert to PP/DS mode. It would also require time-continuous resources that CTM could work with.

This functionality has been active in CTM for some time; however, I have never seen it enabled by a client. It brings up some interesting questions.

1. *Acceptance of the CTM Prioritization? CTM in PP/DS mode would directly prioritize the planned orders based upon the CTM prioritization logic, as explained in these articles: http://www.scmfocus.com/sapplanning/2012/06/05/ctm-customer-priorities-versus-order-priorities/ http://www.scmfocus.com/sapplanning/2009/12/08/customer-prioritization-and-ctm/)*

2. *Would PP/DS Procedure Be Used After the CTM PP/DS Run? CTM can be capacity-constrained, so a CTM run in PP/DS mode would result in a feasible plan, and one for which no conversion of SNP orders to PP/DS orders would be required. However, at this point, would the company want to run any heuristics in PP/DS on these planned orders? It would seem that if a company were dedicated to the CTM prioritization sequence, that the next step would be to manually manipulate the CTM-generated planned orders in the PP/DS detailed scheduling board.*

3. *What Would be the Duration of the CTM Production Planning Horizon? In most cases a production planning horizon is somewhere between two and four weeks. However, in this case, the production planning horizon could be as long as the supply planning horizon. In addition, there would be no purpose in setting a value for the production planning horizon, as there would be no overlap between the supply planning horizon and the production planning horizon (as is normally the case when SNP and PP/DS are co-implemented). Instead, there would be a single combined supply and production planning run. I could see (as is stated in the bullet point above) that PP/DS is only used to make manual changes through the PP/DS detailed scheduling board. That would be one design. Another design could have PP/DS heuristics used, and in that case, the supply versus production planning horizon and the overlap would again be an issue.*

4. *CTM may be able to work in a bucket-time orientation, and can create PP/DS orders; however, it cannot do things like incorporate a set-up matrix, something that PP/DS can do (see this article on the changeover planning: http://www.scmfocus.com/productionplanningandscheduling/2010/12/06/changeover-planning-in-sap-ppds-vs-planettogether/). Therefore, CTM cannot be as accurate as PP/DS. On the other hand, most companies that I have worked with have problems effectively implementing the set-up matrix in APO, and may fall back to emulating the set-up*

times with a production cycle, so this would seem to be a more hypothetical than real problem. For companies that want a solution that is effective at managing setup/changeover times, there are much better solutions than PP/DS.

The Conversion of SNP Planned Orders to PP/DS Planned Orders

Once the SNP planned orders are converted into PP/DS orders, they are released for sequencing. Once the sequencing is completed, the PP/DS orders are released for production. Ordinarily, the SNP orders must be converted to PP/DS orders. Time-continuous planning is applied to PP/DS PPMs and PDSs and provides the highest detail available within APO. Bucket-oriented planning on the other hand uses the SNP PPM or PDS, which is more abstract. The system can automatically derive the bucket capacity of a mixed resource from the time-continuous capacity. Since you do not plan with so much detail in SNP, for example you do not use sequence-dependent setup times, you can reduce the bucket capacity derived using a loss factor. You obtain such a buffer for detailed planning in PP/DS.

SAP help has the following comments on SNP planned order conversion.

> *SNP orders converted to PP/DS orders are still visible in SNP as aggregated demands. If you use the same resources in SNP and in PP/DS (mixed resources), the resource schedule is visible in SNP and in PP/DS.*

Converting SNP orders individually	*From the Interactive Production Planning menu area, access the product view for the product. Call the order processing view for the receipt element to be converted. Choose Edit → Convert SNP order.*
Converting all SNP orders in the production horizon	*From the Production Planning menu area, choose Environment → Conversion of Supply Network Planning → Production Planning.*

When performing the second option, the following screen will appear.

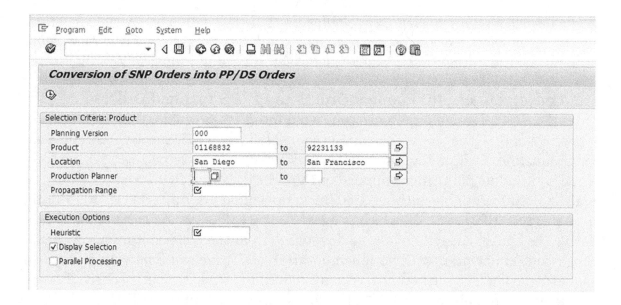

The Conversion of SNP Planned Orders to PP/DS Planned Orders When CTM is Used in PP/DS Mode

The PP/DS orders that are created by CTM can be used without conversion when time-continuous planning is used in CTM, as described in the quotation below:

> *Mixed resources have both the SNP bucket and PP/DS time-continuous capacity definition. CTM planning can use mixed resources for planning. When planning in PP/DS mode, CTM uses the time-continuous capacity as the primary capacity for finite planning. The bucket capacity is calculated for the scheduled activity to keep the time-continuous and bucket-continuous capacity requirement consistent. On the other hand, when planning in SNP mode using mixed resources, the bucket capacity is used for finite planning, but time-continuous capacity isn't calculated. –**Capable to Match (CTM) with SAP APO***

Therefore, if a CTM Profile is required to process in time-continuous mode for a month or two, and then process in bucket-oriented planning mode, it often makes sense to break up the planning horizon with two CTM Profiles that are identical except for the time orientation and which PPMs and PDSs are used (SNP for bucket-oriented and PP/DS for time-continuous).

When it comes to conversion of PP/DS planned orders to production orders, in most cases SNP orders have a conversion indicator for the planned orders that can be set manually by the production planner. The relevant planned orders are automatically converted into production orders and are triggered by the conversion in APO.

Let us evaluate each of the nonproduction constraint types.

Storage Constraints

While it is not the only way to set up a storage constraint, the most common way that I have seen is to set up one storage constraint for the entire location. The unit of measure used for the storage capacity is quite flexible. A location's capacity could be stated in terms of cases that can be accepted by a facility (e.g., 1.5 million cases), or it could be stated in pallet spots. However, in many cases it does not make a lot of sense to have such a high level constraint as the number of cases because the product may need to be in a particular place rather than stored "anywhere." What this means is that a location could be under-capacity regarding the aggregate resource that is stated in cases, but over-capacity for a particular area within the location. It would be necessary to place more storage resources into the system so that the location could be modeled more accurately.

However, it turns out that storage resources are far more "elastic" than production resources. Extra storage is usually available in the form of either third-party warehouses or trailers in the yard. So the total capacity within the four walls is not necessarily limited to the storage capacity at the location's physical building. I have found on several occasions that the actual business requirement vis-à-vis storage resources were the capacity of locations to ship out or receive in certain quantities of product. This would be best modeled with a handling constraint, which I will discuss next. A handling constraint is completely different from a storage constraint at a location.

Handling Resources and Location-to-Location Flow Constraints

If you analyze the constraints related to storage, it is sometimes the case that shipping and receiving are the actual constraints, or what would translate into the handling resource. Locations can only ship and receive a certain amount per day. Without a constraint in place, some days the supply planning system will schedule more volume than the facility can process, meaning that the volume must be pushed manually to the next day.

Setting up a handling resource and constraining the resource for product both arriving and departing from the location can meet this requirement. The maximum amount of product that could be shipped or received in a given day per location would thus be capped. Essentially, a daily aggregate capacity constraint can be accomplished by setting one inbound handling resource to finite and one outbound handling resource to finite.

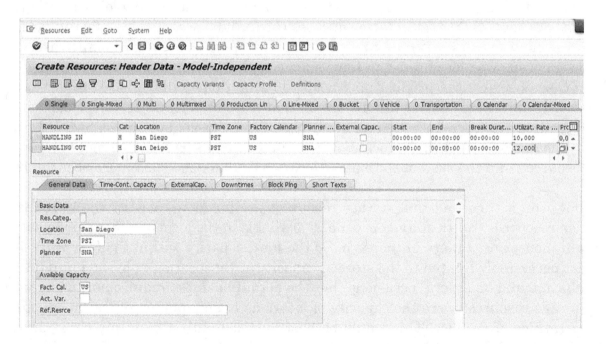

This is in fact the only scenario—outside of some specialty product—which requires a unique handling resource, where a handling resource for the purposes of constraining capacity would occur. Using a handling resource as described here actually makes a lot of sense, and while I would like to see it implemented, I have yet to see this happen at any company.

Transportation Constraints

For most companies that ship product, transportation constraints for supply planning are not very useful, as most companies can obtain extra transportation capacity when needed. Transportation constraints are different from transportation optimizers for vehicle scheduling. Transportation optimizers are quite useful in optimizing the available time of a truck to get the maximum freight delivered to the most locations. The article below describes a particularly effective and quite inexpensive vehicle scheduling application.

http://www.scmfocus.com/supplychaininnovation/2009/06/google-maps-and-gomobileiq-for-vehicle-routing/

Instead, I am referring to the use of transportation resources in a supply planning application such as SNP. Someone might bring up the topic of private fleets and whether companies that use private fleets may benefit from the use of transportation resources in their supply planning application. It's not that companies don't use private fleets, but in the vast majority of cases, the private fleet is augmented with public for-hire carriers (as is discussed in the article below). Companies that use private fleets one hundred percent of the time, and do not augment their private fleet with outsourced transportation, are exceedingly rare.

http://www.scmfocus.com/fourthpartylogistics/2012/04/the-overestimation-of-outsourced-logistics/

Most companies prefer some carriers over other carriers, and on occasion the preferred carriers are unable to provide capacity. The question is not so much general capacity, but the lead-time between when the request is made for transportation services and when the services can be provided. A good planning system should allow the company to provide adequate lead time on transportation requests, and by placing transportation resources in supply planning applications, a company could get advanced warning when extra external transportation services would be required. But this does not seem to be sufficient incentive to get companies to model their transportation resources.

Resource Uptime

The utilization on the resource can be reduced to equal the average of the change-over time for that resource. This would reduce the capacity of the resource to reflect downtime between changeovers (for SNP).

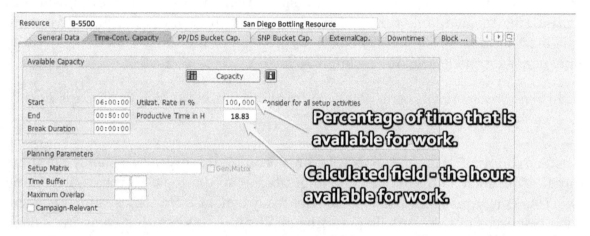

The utilization can be adjusted here in the resource. The utilization adjustment can be used to account for the percentage of time that the resource is not available to do work. The utilization should always be below one hundred percent. Some companies simply take the number of hours that the resource (or production line), was available for work the previous year, and then divide that by the resource's capacity to arrive at a percentage.

Therefore, using a period lot size/manufacturing cycle is a more aggregated and less accurate way of keeping manufacturing efficiency high than using a change-over matrix.

BOMs, Routings and Work Centers and the PPM and PDS

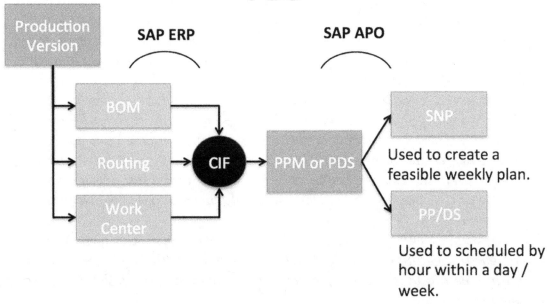

SNP uses the same resources as are used in PP/DS, but in a more aggregated way. What this means is that SNP is "aware" of the same resources that are used in PP/DS. While SNP can use different resources than are used in PP/DS, this is generally not a very good design, and connecting the systems would be higher in maintenance and confusing to manage. Therefore, it is best to use the same production resources in SNP and PP/DS, although resources can be aggregated using the aggregated resource functionality in SNP (although in practice this is done rarely). I do not cover aggregate resource functionality in this book.

Bottleneck Resources in Production Planning

The theory of bottleneck resources is one of the most enduring in production scheduling. I quote from *The Encyclopedia of Operations Management*:

> *Theory of Constraints (TOC)—A management philosophy developed by Dr. Eliyahu M. Goldratt and popularized by his book The Goal, which focuses on the bottleneck resources to improve overall system performance. The TOC recognizes that an organization usually has just one resource that defines its capacity. Goldratt argues that all systems are constrained by one and only one resource. As Goldratt states "a chain is only as strong as its weakest link." This is an application of Pareto's Law to process management and process improvement. TOC concepts are consistent with managerial economics that teach that the setup cost of a bottleneck resource is the opportunity cost of the lost gross margin and that the opportunity cost for a non-bottleneck resource is nearly zero.*

> *According to TOC, the overall performance of a system can be improved when an organization identifies its constraint (the bottleneck) and manages the bottleneck effectively. TOC promotes the following five-step methodology:*

> *1. Identify the System Constraint: Finding the largest queue can often discover the constraint.*

> *2. Exploit the System Constraints: Protect the bottleneck constraint so that no capacity is wasted.*

> *3. Subordinate Everything Else to the System Constraint: Ensure that all other resources (the unconstrained resources) support the system constraint, even if it reduces the efficiency of these resources. For example, unconstrained resources can produce smaller lot sizes so the constrained resource is never starved.*

> *4. Elevate the System Constraints: If this resource is still a constraint, find more capacity. More capacity can be found by working additional hours, using alternate routings, purchasing capital equipment, or subcontracting.*

   5. *Go Back to Step 1: After a constraint problem is solved, go back to the beginning and start over. This is a continuous process of improvement.*

Of course, TOC is a process for improvement. Setting up a bottleneck constraint in an advanced planning application is one particular method or sub-approach of managing the bottleneck resource. Some of the recommendations in the methodology above would occur before the bottleneck is set up in APO. For instance, if a company subcontracts a bottleneck resource, then it may or may not also be entered as a bottleneck resource in SAP APO. That is, the act of subcontracting may mean that the resource is no longer the bottleneck.

It is important to review the theory of constraints, because when SAP was developing SNP and PP/DS it was at least in part copying best-of-breed vendors like i2 Technologies that based their design on TOC. (I have yet to come across any functionality in APO that did not already exist in a best-of-breed vendor.) TOC has died down a bit, but at one time Goldratt's book was extremely influential among manufacturing companies and software vendors that made manufacturing software. Think of the mania surrounding Six Sigma today; that was the popularity level of TOC (roughly speaking) in the 1990s.

Background on Bottleneck Resources

To begin this section, I have included the following definition of a bottleneck resource from Wikipedia:

> *A bottleneck is a phenomenon where the performance or capacity of an entire system is limited by a single or limited number of components or resources. The term bottleneck is taken from the 'assets are water' metaphor. As water is poured out of a bottle, the rate of outflow is limited by the width of the conduit of exit—that is, bottleneck. By increasing the width of the bottleneck one can increase the rate at which the water flows out of the neck at different frequencies. Such limiting components of a system are sometimes referred to as bottleneck points.*

As I described in a previous section, the concept of a bottleneck resource is more applicable to production resources than any of the supply planning resources (storage, handling unit, transportation etc.). The main concept behind bottleneck resources is that the bottleneck is the pacemaker for the entire production line or the entire manufacturing process.

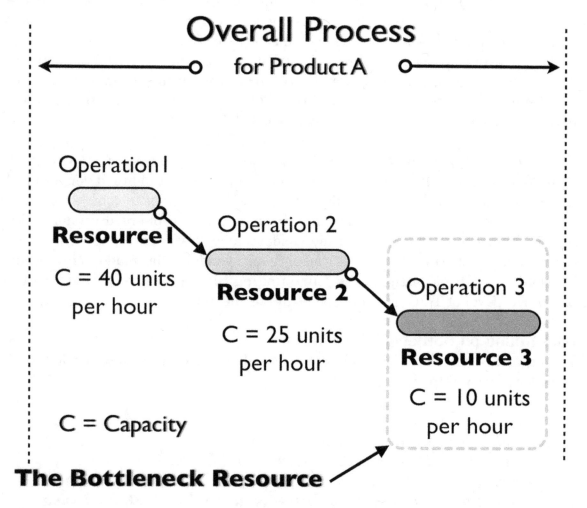

The bottleneck operation and resource is Operation Three. Because Resource Three produces the lowest number of units per hour, it restricts the capacity of the overall process, or sequence, to ten units per hour. The overall process cannot exceed the capacity of the slowest resource (unless the slowest resource runs for more hours per day than the rest of the resources). For this reason, it is unnecessary to constrain the capacity of either Resource 1 or Resource 2. This is the standard example and typically applies to a manufacturing problem.

However, while constraint-based planning works better for manufacturing than for supply planning resources, the above scenario works better for **some manufacturing environments than others**. Very rarely is this fact regarding the match between the application and the environment brought up on projects and especially during sales cycles. Furthermore, some vendors are more limited in the manufacturing environments than they can effectively model. The scenario above works best for discrete manufacturing, where the manufacturing process is a straight "line" with a work center/resource directly transferring an item to the next work center/resource and then directly to the next work center/resource and so on until the production process is complete. It does not work as well when there are leading and lagging operations (for instance, on a product that is incrementally processed), because a lagged operation can process part of the upstream work in progress and is not dependent upon the completion of the previous workstation to begin its work.

There is a strong tendency to over-generalize discrete manufacturing environments to all manufacturing environments, and this over-generalization is not only restricted to software. There are many different manufacturing environments: discrete, process complex, process simple (i.e., mixing operations), repetitive, job shop (i.e., make to order). When permutations are included, the list grows. Nowhere is this overgeneralization more pronounced than in the overemphasis on the Toyota manufacturing system—often referred to as "lean." Toyota's system is a discrete manufacturing system, and therefore its applicability is limited outside of that manufacturing environment.

http://www.scmfocus.com/productionplanningandscheduling/2012/08/25/the-over-generalization-of-discrete-manufacturing-inventory-management/

The book *Production Planning in SAP APO* makes the following point regarding lean manufacturing and the production planning method employed.

This section considers a scenario where there is a bottleneck resource and in which the setup problem is of lesser significance... (this statement describes a manufacturing environment where changeovers are of minimal planning importance) *...the scenarios mapped must not be* *too complex. Such cases are especially widespread in the area of Lean manufacturing. Here the topic of setting up usually plays a secondary role, because planned orders are "long-running" or many similar planned orders are scheduled in a row. In such cases, the mapping of a setup activity is performed, if necessary, by simply blocking the resource using an appropriate dimensionally planned order that is scheduled for a dummy material that has previously been created for this.*

The end of this quotation gets a bit complex and trails off into a solution that is outside of our focus. The interesting thing about the quote is that manufacturing environments, which do not meet the criteria listed above, are frequently the recipient of "Lean manufacturing" initiatives. I have noticed this as well in my consulting work. However, manufacturing environments that are appropriate for Lean manufacturing are specific and can be categorized. Unfortunately, many companies are using Lean-manufacturing principles on environments for which Lean is a poor fit. However, the literature on Lean does not really explain that Lean is designed for specific environments. The way that the literature treats Lean is similar to a pharmaceutical advertisement that makes a drug approved by the Food and Drug Administration for a narrow application appear applicable to the widest possible audience. Most material about Lean proposes that it is a philosophy that applies to every manufacturing environment and also supply planning. There is no attempt to communicate any nuances. For instance, the presentation delivered at ASUG entitled *Enabling Lean Supply Chain Planning* is symptomatic of the material that overgeneralizes and is highly promotional rather than scientifically oriented. Quotes like the following are quite common:

> *Lean principles can have huge returns in reduced inventory carrying cost, scrap and obsolescence.*

Like most promotional literature, there is very little emphasis on the evidence and applicability of an approach, and it is expected that the audience will accept the material uncritically.[7] According to the promotional material, some company somewhere benefitted from the approach at some point in time, so now it's time to do the same thing in your company. Before applicability is established, the typical presentation quickly jumps into the details of how to implement the approach.

Bottleneck Resources in SAP APO

Now that we have covered the definition and etymology of bottleneck resources, we can move on to the topic of the bottleneck resource functionality in SAP APO. There is no direct translation of the general term "bottleneck resource" (as we have just discussed it) to SAP APO. SAP Help states that the following functions apply to a bottleneck resource:

1. *Different Viewing:* On the detailed scheduling planning board (the main user interface for scheduling in PP/DS), bottleneck resources can be displayed differently than non-bottleneck resources.

2. *Separate Campaign Optimization:* In the optimization function of LS>Production Planning and Detailed Scheduling, you can carry out a campaign optimization for bottleneck resources.

The first item shows how the resource is displayed in the interface. The second item shows how the bottleneck resources can be optimized separately from the other resources. However, in both cases this is identification functionality, not any extra functionality that can do something differently with a bottleneck resource that cannot be done to any ordinary finite resource. This is strange because the application definitely gives the impression that it offers some special functionality specific to "bottlenecks."

---

[7] Many of these references to a well-known company are highly suspect. I have now observed several examples of references to a company receiving some extraordinary benefit by implementing some supply chain planning approach, only to have worked in that company and found that they did not do what they were referenced to have done. Many consulting companies simply make up stories about benefits that never happened. Simply having the actual client co-present at a conference with the consulting company is not evidence that the story occurred as they said it did. I have worked on projects where the client had yet to see the benefits proposed by the consulting firm, and where the client was quite unhappy with the performance of the approach and the consulting company. Yet the client representatives still attended the conference with their consulting company and put on a brave face. I have found that companies have the same incentive to appear successful as the consulting companies have in presenting what are often illusory success stories.

It bears mentioning that bottleneck resource functionality is entirely in PP/DS and does not extend to SNP; without PP/DS this functionality cannot be used. SNP creates an initial supply and production plan—which is not aware of the bottleneck designation in the resource that is applied in PP/DS. APO then requires that PP/DS perform its own optimization on bottleneck resources—which is likely to change the initial supply plan created by SNP—and then send this back to SNP. The optimizer screen shows where this is set up.

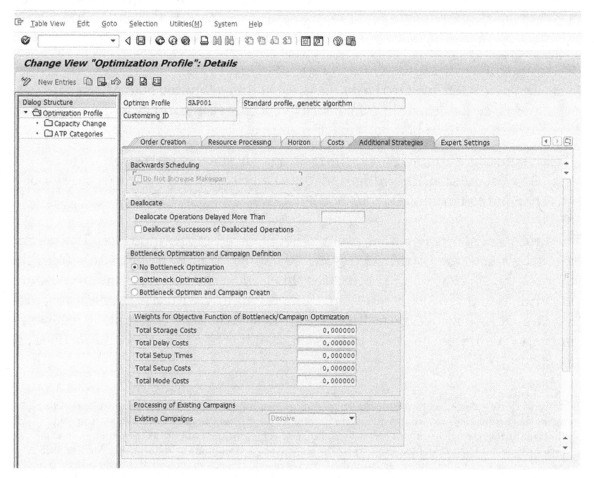

The PP/DS optimizer has the option of paying no attention to the bottleneck setting on the resource, or of constraining all bottleneck resources.

How a bottleneck resource is integrated with infinite resources is an interesting topic, but one which can be difficult to get a solid grasp on.

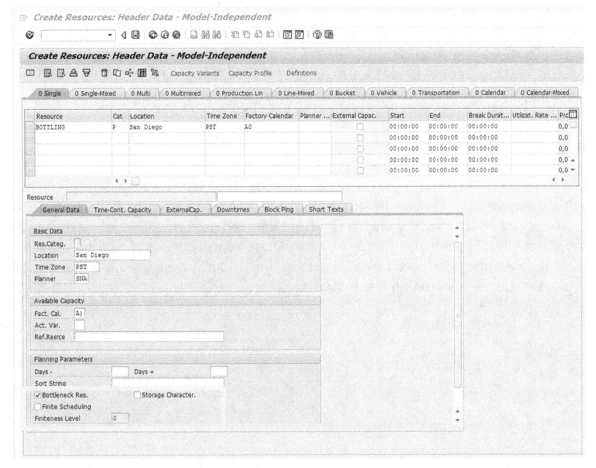

As the screen shot above shows, a resource can be set as infinite, or a bottleneck, or both.

While using a resource in a finite manner has only a few prerequisites (the finite toggle switch on the resource must be selected and the constraint-based method used must be set to finite), using a bottleneck resource is more involved.

Therefore, APO's bottleneck functionality can really be described as "identification functionality." By setting the bottleneck flag on the resource, it can be viewed and can be made part of a campaign optimization. While I have not seen this used on a project, according to the book *Production Planning in SAP APO*, this is commonly employed in the chemical and pharmaceutical industries.

However, SAP's bottleneck functionality and description is inconsistent with the actual definition of a bottleneck resource. A company performs bottleneck resource planning if they take the following steps:

1. The bottleneck resource setting is not checked for any resource.

2. The finite setting is set for a single resource on a production line.

3. The CTM or the SNP or PP/DS Optimizer is run.

If the company also decides to set the bottleneck setting for some or all of the finite resources, then they will be able to identify them and perform a special optimization for them—but this should simply be considered extra functionality. The resource actually becomes the bottleneck resource—or co-bottleneck resource—once the finite setting is selected on the resource.

When Multiple Resources Along a Production Line Must be Modeled

When a manufacturing process is sequential and straightforward to model, it can be finitely planned on the basis of the bottleneck resource. In that circumstance, selecting the right resource to set as the bottleneck resource is simple.

The Production and Resource Scenarios

In this section I will show a variety of production requirements that companies face in real life. The first few scenarios are included to explain the basics of how the resources work in APO, and are too simple to represent real life scenarios. However, as we move through the different scenarios, they will become more and more complex and realistic. With each new scenario we will touch on issues that companies face in modeling their production resources, and we will discuss modeling approaches for addressing these requirements.

Most often many resources are involved in a production process. However, I model only two resources in these examples because I can explain what I need to with just two resources, making it is easier for the reader to follow. In general, production planning does not focus attention onto every resource along a production process, but to just one or two resources. However, I will show in a few of these scenarios that constraining one resource would be insufficient to model the overall production line. Below each graphic I have included the salient features of the scenario.

1. Sequential Resource Constraints

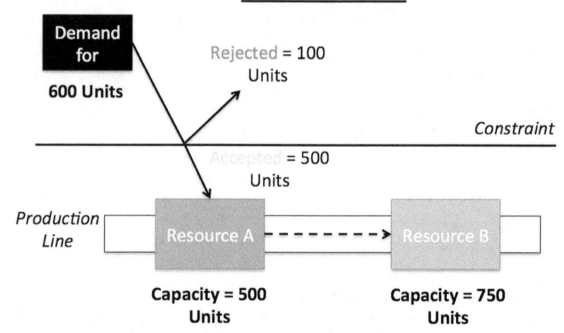

Salient Scenario Features

1. *Finite Resource?—**A***

2. *Resources Feeding or Being Fed by Multiple Resources?—**No***

3. *Different Capacity Per Product Per Resource?—**No***

4. *Could Some of the Resources be Aggregated?—**N/A***

5. *Aggregation Reason?—**N/A***

This is the simplest possible design. The graphic simply explains that if a resource is set as finite, then it restricts the demand that can be accepted/scheduled for the entire production line. That single resource restricts the capacity for the entire production line, and it does not matter if the resource is the first or the last in the production sequence. Also, "rejected" is a general term to mean "not accepted at that period." Automatic capacity leveling that is part of constraint-based planning will attempt to move the demand forward or backward (depending upon the backward and forward scheduling configuration).[8]

The larger capacity of resource B is immaterial to the production rate of this line because the bottleneck resource and where the constraint is placed—resource A—is the pace-setter, and restricts the quantity that can be produced to 500 units. The extra 250 unit capacity of resource B simply goes to waste.

I sometimes get questions about what happens if the resources are set to different units of measure. For instance, in beverage production, the liquid processing resource is measured in gallons per hour, while the bottling resource is measured in bottles per hour, and the final demand is in bottles per hour.

However, this does not mean that the bottleneck resource needs to be set at the bottling resource, if the liquid processing resource is actually the bottleneck. APO will convert all of the capacities and units of measure between the resources. Imagine the following scenario which illustrates how this works:

---

[8] I include forward and backward scheduling in my book *Planning Horizons, Calendars and Timing in SAP APO*. The book compares and contrasts forward and backward scheduling with forward and backward forecast consumption—two areas of functionality that can be easy to confuse with one another.

2. Different Units of Measure Per Resource

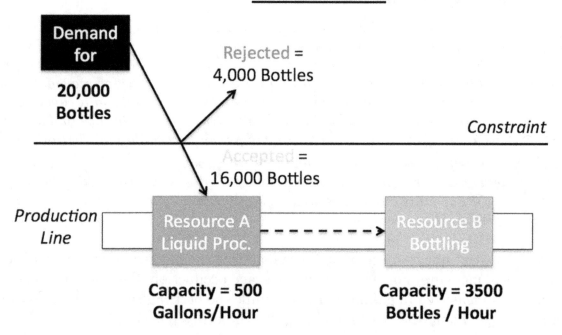

Available Time for Processing = 8 Hours

Salient Scenario Features

1. *Finite Resource?—**A***

2. *Resources Feeding or Being Fed by Multiple Resources?—**No***

3. *Different Capacity Per Product Per Resource?—**No***

4. *Could Some of the Resources be Aggregated?—**No***

5. *Aggregation Reason?—**N/A***

How does the system know that it should reject 4,000 bottles and accept and create a planned order for 16,000 bottles? One of the functions of the planning system is that it does all of the math that I have listed above automatically. This second scenario is very similar to the first, except that it adjusts for different units of measure between the liquid processing resource and the bottling resource. SAP APO converts the units of measure across a series of resources on a production line. This is standard within the product, and as long as the master data for the product and the resource is set up accurately, APO will perform all of the conversion mathematics. All the company has to do is enter accurate master data and the system will manage the rest.

Constraint Based Planning Math

Requirement	Units
Demand in Bottles	20,000
Resource Capacity in Gallons Per Hour	500
Available Hours on Liquid Processing Resource	8
Resource Capacity in 8 Hours in Gallons	4,000
Number of Bottles Per Gallon	4
Gallons in Bottles That Can be Produced in the Available Time on the Liquid Processing Resource	16,000
Demand That Can be Accepted and Planned	16,000
Demand That Must be Rejected to Moved to a New Period	4,000

Although the ability to perform all of the unit of measure conversion mathematics is infrequently discussed (probably because it is seen as basic functionality), it is a major productivity enhancement with supply planning and production planning systems. Imagine performing these calculations by hand, the potential errors that could ensue, and the costs of those errors? It was not that long ago that all production planning unit of measure conversions were calculated this way.[9]

---

[9] Collectively, we don't spend much time studying the history of supply chain planning. Most works of history instead focus on politics, personalities, and warfare with very light coverage given to the much less glamorous aspects of human activity. I have searched and have been unable to find books (not more than a few actually) or universities that concentrate on this topic. However, it is interesting to learn how things were done in all aspects of supply chain planning, particularly prior to computerization. I have learned a great deal by studying this area, and in particular how incorrect many projections are from those who were ostensibly "experts" in this area. I was always suspicious of the claim that Henry Ford invented the assembly line (in truth, outside of MBA courses, historians only credit him with the automation of the assembly line rather than its actual invention). However the first assembly line for which I could find evidence was The Venetian Arsenal, which built ships for the Venetian city-state over a thousand years ago and was in continual operation for eight hundred years. This topic is covered in the following article: http://www.scmfocus.com/productionplanningandscheduling/2012/07/04/who-was-the-first-to-engage-in-mass-production-ford-or-the-venetians/
   http://www.scmfocus.com/scmhistory/

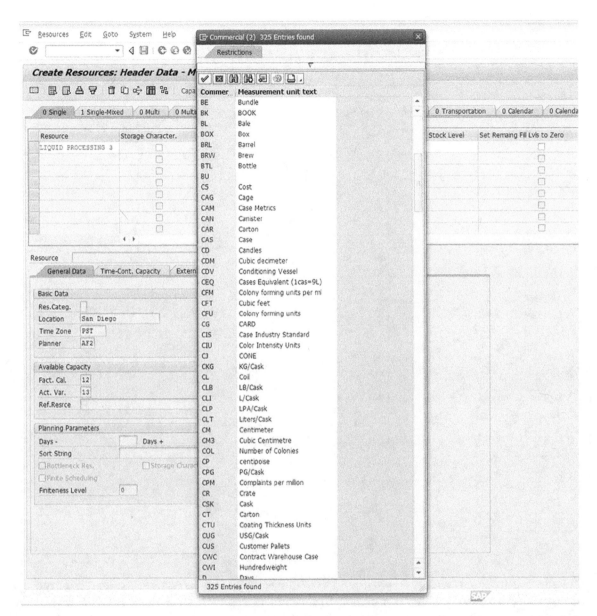

The resources in APO can support an enormous number of units of measure as can be seen from this screen shot. This drop down shows the three hundred and twenty-five measurement units that are available to choose from.

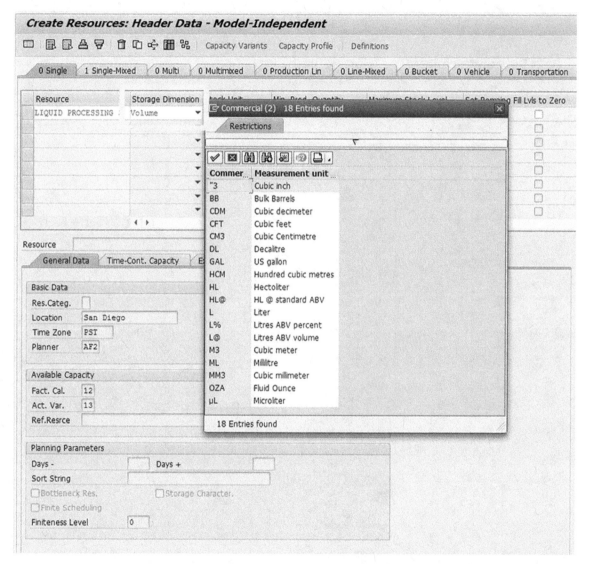

However, the "storage dimension" which can be seen in the screen shot above categorizes the units of measure. Once a storage dimension is selected, the stock units that can be selected are restricted to those that are valid for the storage dimension. In this case, because volume was selected as the storage dimension, the stock units are now restricted to just 18.

3. Multiple Resources Per Area: Not Interchangeable

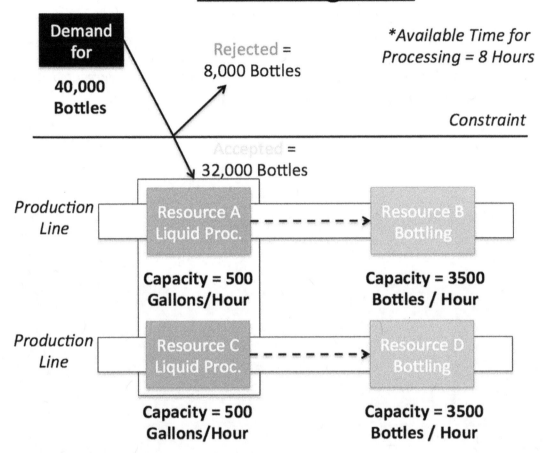

Salient Scenario Features

1. *Finite Resource?—**A, C***

2. *Resources Feeding or Being Fed by Multiple Resources?—**No***

3. *Different Capacity Per Product Per Resource?—**No***

4. *Could Some of the Resources be Aggregated?—**No***

5. *Aggregation Reason?—**N/A***

With this scenario we begin to get a bit more realistic. We have two parallel production lines. They have the same capacity per resource, and they do not differ in rates per product mix. So nothing much has changed from the previous scenario except that we have two production lines instead of one. Due to the physical setup of the plant, resource A can only feed resource B, and resource C can only feed resource D.

This example would be a prime example of an opportunity for resource aggregation. Resources need only be treated as separate if they differ in some way. If not, resources can and should be combined or aggregated. This is different from the aggregated resource functionality in SNP.

4. Multiple Resources Per Area: Interchangeable: Same Capacity Per Product Per Resource

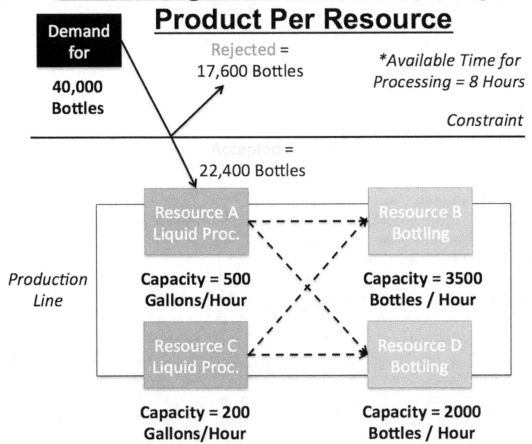

Demand for

40,000 Bottles

Rejected = 17,600 Bottles

Available Time for Processing = 8 Hours

Constraint

Accepted = 22,400 Bottles

Production Line

Resource A Liquid Proc.

Capacity = 500 Gallons/Hour

Resource B Bottling

Capacity = 3500 Bottles / Hour

Resource C Liquid Proc.

Capacity = 200 Gallons/Hour

Resource D Bottling

Capacity = 2000 Bottles / Hour

Salient Scenario Features

1. *Finite Resource?—**A, C***

2. *Resources Feeding or Being Fed by Multiple Resources?—**Yes***

3. *Different Capacity Per Product Per Resource?—**No***

4. *Could Some of the Resources be Aggregated?—**Yes (A & C), (B & D)***

5. *Aggregation Reason?—**Both A & C and B & D could be aggregated because they do the same processing and do not differ in their rates based upon products processed. Also, the resources are now interchangeable.***

Several changes dramatically alter this scenario from the previous scenario. First, resources A and C can now both feed resources B or D. Any product can be processed through any resource, enabling the two separate production lines to be treated as one large production line.

A second change from the previous scenario is that the capacities of the liquid resources (A & C) and the bottling resources (B & D) are no longer the same. However, this is not material to whether the resources can be aggregated. Resource AC would be set up in the system with a capacity of 700 gallons per hour and resource BD would be set up with a capacity of 5500 bottles per hour.

Multiple Resource Constraints

Up to this point the scenarios have focused only on the first resource, the liquid processing resource, as being a constraint to the production process. In the following scenarios we will add an extra complexity, which is based upon the effect on capacity of changing the product mix. Here the capacity of the bottling line changes depending upon which product is to be bottled. This scenario is actually very easy to visualize. Simply increasing the size of the bottle would tend to increase the capacity of the bottling operations, because larger bottles have a smaller relative time in motion through the bottling resource versus the volume of liquid processed. Having worked for several beverage companies, I can say that in practice there are quite a few reasons as to why a bottling line capacity could increase per product being processed. However, let's focus on bottle size as that is the easiest to understand. What this means is that if we apply the same modeling that we have up to this point, one particular product mix (which would have larger bottles) would make the liquid processing resource the constraint. However, in other situations, with a different product mix that has smaller bottles, the bottleneck resource would become the bottling resource (no pun intended).

So how can this scenario be properly modeled? One might say that both resources could be set to finite or constrained. However, would that work? Multi-constrained resources can be set for a production process. Therefore the optimization must respect both the liquid constraint and the bottling constraint. So if the bottling constraint is reached before the liquid constraint (because the liquid capacity is higher than the bottling constraint), the optimization must create a planned order for what can be processed at the bottling constraint only, even though there is excess capacity on the liquid constraint. Multiple finite resource planning is standard functionality in both SNP and PP/DS. The book *Production Planning in SAP APO* states it the following way:

> *You can use the PP/DS Optimizer to take several bottlenecks into account at the same time, order priorities can be included, and a multilevel optimization is possible. Because the PP/DS Optimizer does not create or delete any orders, but merely reschedules them, a previous requirements planning is a prerequisite for using the PP/DS Optimizer.*

5a1. Multiple Resources Per Area: Interchangeable: Same Capacity for All Products / Liquid and Bottling Constraints

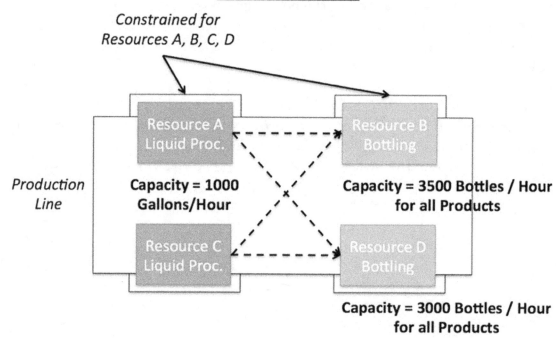

Salient Scenario Features

1. *Finite Resource?*—**A-C, B-D**

2. *Resources Feeding or Being Fed by Multiple Resources?*—**Yes**

3. *Different Capacity Per Product Per Resource?*—**No**

4. *Could Some of the Resources be Aggregated?*—**Yes (A & C), (B & D)**

5. *Aggregation Reason?*—**Both A & C and B & D could be aggregated because they do the same processing and do not differ in their rates based upon products processed. Also, the resources are now interchangeable.**

This scenario differs from scenario 4 because of the dual constraint, with the bottling resource being an additional constraint; prior to this scenario, only liquid has been a constraint.

5a2. Multiple Resources Per Area: Interchangeable: Different Capacity for All Products / Liquid and Bottling Constraints

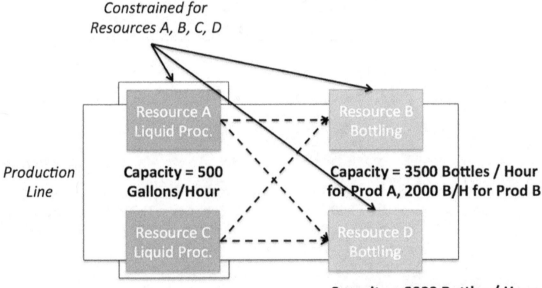

Salient Scenario Features

1. *Finite Resource?*—**A–C, B, D**

2. *Resources Feeding or Being Fed by Multiple Resources?*—**Yes**

3. *Different Capacity Per Product Per Resource?*—**Yes (for B & D only)**

4. *Could Some of the Resources be Aggregated?*—**Yes (A & C)**

5. *Aggregation Reason?—**Resource A & C could be aggregated because they do the same processing and do not differ in their rates based upon products processed. But resource B & D could not.***

The difference between this scenario and scenario 5a1 is that there are now different capacities on the bottling line depending upon the product mix (in this case bigger bottles versus smaller bottles).

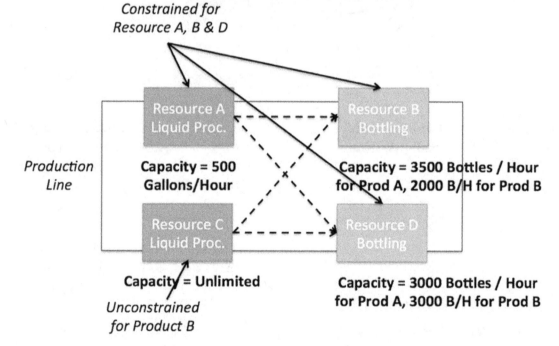

5b. Multiple Resources Per Area: Interchangeable: Different Capacity for All Products / One Liquid and Two Bottling Constraints

Salient Scenario Features

1. *Finite Resource?—**A, B, D***

2. *Resources Feeding or Being Fed by Multiple Resources?—**Yes***

3. *Different Capacity Per Product Per Resource?—**Yes (for B & D only)***

4. *Could Some of the Resources be Aggregated?—**No***

5. *Aggregation Reason?—**Resources cannot be aggregated because of different capacities per resources per product, and because one liquid resource is unconstrained (because it has unlimited capacity for product B, while the other bottling resource is constrained.***

The difference between this scenario and scenario 5a2 is that one liquid resource is unconstrained for one product. The other resource is constrained for all products. (We only assign product B to the unconstrained resource C, and C is not assigned to resource A.)

Hard versus Soft Constraints

In one dimension any optimizer can treat constraints in different ways. There are two extreme forms of constraints, which also happen to be the two most commonly used resource types: hard constraints and soft constraints. Resources are an example of a hard constraint; the optimizer may not exceed or "violate" this type of constraint under any circumstance. Hard constraints make a lot of sense because a production resource that can produce one thousand units per day cannot be made to exceed that production rate without incurring substantially more costs. Hard constraints do not require costs to be associated with them because there is no option to exceed their stated capacities. Things such as transportation costs are not constraints; these costs simply scale with the transportation units scheduled. However, there can be transportation resources, and the transportation resource capacity cannot be exceeded when they are used in their default mode of hard constraint.

A constraint may also be treated as soft; here the constraint may be violated, but at a cost—most often referred to as a penalty or penalty cost. Two of the most important penalty costs/soft constraints in SNP are the penalty cost for violating the safety stock target and the non-delivery penalty cost (or the activity of failing to meet a demand). To see how the safety stock penalty works, here is an example: If a safety stock target is 10 units, and the company holds only 8 units for one planning bucket, the optimizer incurs 10-8 = 2, which is The Penalty Cost for that planning bucket. The safety stock penalty cost is incurred in every bucket for which the safety stock target is violated.

Some costs, such as storage costs or transportation costs, are positive costs in that the optimizer incurs them at the time of activity. Penalty costs are negative costs, and are incurred when the optimizer fails to do something that is desirable. The same dynamic applies for storage costs. In SNP, storage costs are incurred on a daily basis per unit. Another example of a soft constraint is the cost of unfulfilled demand. The objective function of SNP is to minimize costs, so the unfulfilled demand is declared as a penalty cost to encourage or direct the optimizer to meet as many demands as is feasible, given the hard constraints.

Soft constraints were one of the important conceptual developments in optimization, because they allowed the introduction of implicit costs into the model. While not often discussed, soft constraints are interesting because they allow the combination of explicit (or "real") costs along with "implicit" costs, which are essentially imaginary. Luckily for those of us working in supply chain planning optimization, imaginary costs have been extremely well-developed as a concept and explained in the field of economics where they are under the category of "opportunity costs." Wikipedia describes opportunity costs as follows:

> *Opportunity cost is the cost of any activity measured in terms of the value of the next best alternative forgone (that is not chosen). It is the sacrifice related to the second best choice available to someone, or group, who has picked among several mutually exclusive choices. The opportunity cost is also the "cost" (as a lost benefit) of the forgone products after making a choice. Opportunity cost is a key concept in economics, and has been described as expressing "the basic relationship between scarcity and choice." The notion of opportunity cost plays a crucial part in ensuring that scarce resources are used efficiently.*

Imagine my surprise when I found that the opportunity cost entry in Wikipedia had a full explanation of implicit and explicit costs.

Explicit Costs

Explicit costs are opportunity costs that involve direct monetary payment by producers. The opportunity cost of the factors of production not already owned by a producer is the price that the producer has to pay for them. For instance, a firm spends $100 on electrical power consumed; their opportunity cost is $100. The firm has sacrificed $100, which could have been spent on other factors of production.

Examples of explicit costs in SNP include storage costs and transportation costs.

Implicit Costs

Implicit costs are the opportunity costs in factors of production that a producer already owns. They are equivalent to what the factors could earn for the firm in alternative uses, either operated within the firm or rented out to other firms. For example, a firm pays $300 a month all year for rent on a warehouse that only holds product for six months each year. The firm could rent the warehouse out for the unused six months, at any price (assuming a year-long lease requirement), and that would be the cost that could be spent on other factors of production.

Examples in SNP of implicit costs include non-delivery penalty costs and safety stock penalty costs.

More on safety stock penalty costs can be found in the following article.

http://www.scmfocus.com/sapplanning/2011/11/05/how-soft-constraints-work-with-soft-constraints-days-supply-and-safety-stock-penalty-costs/

Finiteness Level of Resources

Resources can be set as finite or infinite for different applications. The settings for doing this can be found both in the method used (with CTM and the PP/DS and SNP Optimizer), as well as on the resource itself. In APO this is referred to as the "finiteness level."

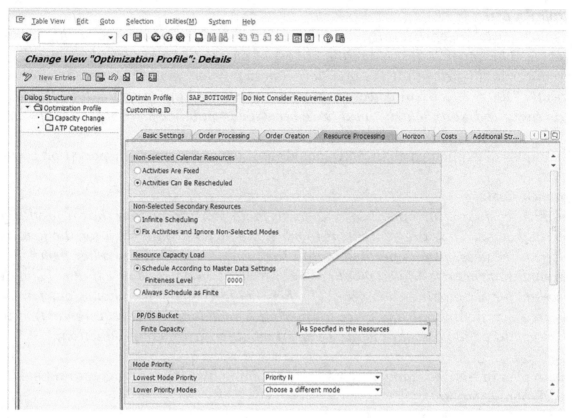

Here we can see the finiteness level setting on the PP/DS Optimizer Profile. In order to know how soft or hard to treat the constraint, the finiteness level must be declared.

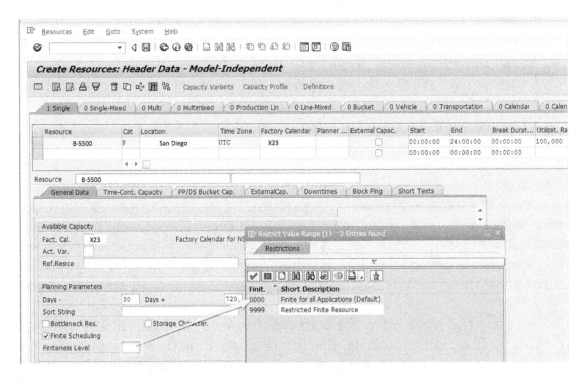

SAP defines the finiteness level of the resource as follows:

*If the resource is used by several applications, you can use the finiteness level to define which application schedules the resource finitely and which schedules infinitely. If you do not assign any finiteness level to the resource, the default value is 0. The maximum possible value is **9999**. If you enter this value, only the applications in which the finiteness level is also **9999** schedule the resource finitely. If you enter the value 0, all applications in which finite scheduling is set schedule these resources finitely, irrespective of the finiteness level defined there.*

Finiteness level	*Short description*
0	*Finite for all applications (default)*
100	*For PP finite resource*
200	*For DS finite resource*
300	*For PP/DS optimization finite resource*
9999	*Restricted finite resource*

In PP the resource should be scheduled infinitely, and in DS it should be scheduled finitely. You enter the finiteness level 200 at the resource R1. In the PP strategy profile, you define that only resources with a finiteness level smaller than or equal to 100 should be scheduled finitely. In the DS strategy profile, you enter 200 for this value. PP therefore schedules infinitely on R1, while DS schedules finitely.

Conclusion

Different domains of the supply chain have different types of resources that must be modeled. For instance, trucks are a resource for supply planning, while a workstation is a resource for production planning. Of the different supply and production planning methods available in APO, only prioritization/allocation (CTM) and cost optimization have the ability to run in a constrained fashion. However, this occurs only if the system is configured to manage resources in a finite manner.

Resources are the mechanism for both constraining the plan in SAP APO and determining if a plan is feasible. Placing resources into a supply planning system is the next step up from using simple lot sizes in the emulation of production constraints. Because of the nature of some manufacturing processes and the fact that they are often sequential, constraint-based applications for production planning leverage this fact into modeling only the bottleneck resource. However, supply planning resources and supply planning in general are less sequential in nature, and some resources (e.g., transportation resources or handling resources) are not constrained in the same manner as production resources. For instance, tendering freight to external providers can easily attain transportation resources. Handling resources are inexpensive and can be easily added. Storage resources, or the capacity in a warehouse, may seem like a hard constraint, but material may be held in trailers outside the facility, or the company may take advantage of third-party warehouses in the case of overflow. Production resources are more valuable to model and to constrain than supply planning resources, as is evidenced by the fact that the most common resources modeled in both supply planning and production planning applications are production resources.

Resources represent the various capacities within both supply and production planning. SNP can use all the resource types (SNP can use production resources in addition to supply planning resources), while PP/DS can only use production resources.

This chapter covered the timing-related fields for resources. With the exception of the bucket resource, most of the timing-related fields across the commonly used resource types are identical. Where they differ is in how the resource is used by APO—whether the resource is a production, handling, storage, or transportation resource. Of the commonly used resource types, the most important distinction between them is the following characteristics:

- Bucket versus Time Continuous

- Mixed versus Non Mixed

- Single (Activity) versus Multi (Activity)

It is the mixed resources that can be used both by SNP and PP/DS, because they contain both timing-related fields for bucket-oriented and time-continuous planning. The most commonly used resources in APO are essentially named as one combination of the characteristics listed above. However, in addition to the resource timings, the method that interacts with these resources can be designed to work in different ways. For instance, CTM—traditionally a supply planning method that works in bucket-oriented planning and generates SNP planned orders—can also be set to time-continuous planning and to generate PP/DS planned orders.

As with a location, a calendar can have a resource associated with it that determines the workdays of the resource. The resource-available times, which are declared within each resource, then operate within the open days of the calendar. A resource is populated with a capacity, which can be constrained or unconstrained, but the location calendar, resource calendar and the start, end, and break duration form another constraint that is as important as the capacity value assigned to the resource.

The resource calendars control when the resource is available for work. First the factory and distribution calendars are set up, and then they are assigned to resources and to planning calendars. The holiday calendar is in turn assigned to the factory and distribution center calendar. Both the holiday calendar and the factory calendar are created using the same screen.

This chapter covered the original theory of bottleneck resources and also how SAP defines "bottleneck" resources. As with many areas of SAP functionality, it turns out that SAP's definition is not the same as the generally accepted definition. The bottleneck functionality in APO is really more about identification and segmentation of resources for special treatment than it is about meeting the classical definition of a bottleneck.

Included in this chapter were a number of scenarios that moved from the most simple to the most complex, and involved the use of two sets of sequential resources: liquid resources and bottling resources. These scenarios showed us how the characteristics of the resources allow them to be set up differently depending upon how the resources process product.

Resources fall into the category of "hard" constraints. Hard constraints cannot be violated at any cost. A second category of constraints—soft constraints—may be violated at a cost. These costs are modeled as penalty costs, which are negative costs incurred when the optimizer fails to do something which would be desirable from the business perspective for the optimizer to do. Soft constraints are interesting because they allow the combination of explicit—or "real"—costs along with "implicit" costs, which are, in a way, imaginary.

In APO a constraint can be set to different levels of "finiteness," which means that the resource can be finite for some applications but infinite for others.

CHAPTER 6

Capacity-constraining Vendors/Suppliers

A number of companies that I have worked with have requested capacity-constraining of suppliers. One can capacity-constrain internal production locations, but to determine overall feasibility it is necessary to capacity-constrain suppliers as well, particularly the largest suppliers. Because of changes to the business environment where manufacturing is increasingly outsourced, the requirement to either constrain or have visibility into suppliers has greatly increased in the past fifteen years.

There are several challenges to capacity-constraining a supplier production resource.

1. *The Technical Challenges of Modeling Supplier Capacity:* APO does allow for suppliers to be modeled, but as more of a workaround, which will be fully covered in this chapter.

2. *The Business Challenges of Modeling Supplier Capacity:* In the early stages of the project, project teams don't often consider how to get suppliers to provide continually-updated capacity information. When a process requires coordination between two companies, the results are uniformly worse than when the

process is managed internally by one company, even though collaboration has been talked up quite a bit. There are many different issues involved in collaboration, including the structure of the relationship between the two collaborating companies, the level of trust, and the incentives to collaborate. Authors of books and articles that have little interest in describing how things work in reality tend to gloss over these issues of collaboration. The fact is that few companies really know how to work collaboratively with other companies. Getting other companies to share their updated capacity information turns out to be a real challenge. The larger a percentage of business the customer company represents, the better the likelihood of getting compliance.

Technical Challenges of Modeling Supplier Capacity

Capacity-constraining suppliers can be both passive and active. Passive constraining is the most basic. Passive constraining tends to be used when the supplier is an "ordinary" supplier. Active constraining is more appropriate when the company is using contract manufacturing or the plant is essentially a "slave plant" to the company running the APO.

Over the last several decades, manufacturing has become intensively outsourced, and manufacturing is viewed less and less as a core competency. The most extreme form of outsourced manufacturing is contract manufacturing. Even companies that are very well known for manufacturing, like Toyota, actually do far less manufacturing than most people realize. The book *Who Really Made Your Car?* explains the high degree to which subcomponents are contract-manufactured in the automotive industry. This book, as well as several other sources, estimate that between twenty-five and thirty percent of the average automobile's value is actually produced by the OEM (Toyota, GM, etc.). The book *Supply Chain Brutalization* points out the following:

> *Automotive provides the volumes of consumer electronics, but contrary to the EMS ODM model of vertical integration, less than 25 percent of an automobile is manufactured by the maker. Components and subsystems, such as ignition systems, tail lights, interior liners, and power*

*seats are all subcontracted to third parties, rendering the car maker to
be a chassis manufacturer and systems integrator, so to speak.*

Who Really Made Your Car? further proposes that one of the major reasons for
Toyota's rise to prominence is not only its highly-touted internal manufacturing processes, but Toyota's ability to serve as a system integrator by developing
highly-collaborative relationships with the many contract manufacturers and
suppliers that make the majority of Toyota's automobiles (with Toyota primarily
performing the final assembly function). For some time at least, Toyota actually
took an in interest in the success of their vendors—a very atypical perspective.

Contract manufacturing is explained in the following quote from *Beyond BOM
101: Next Generation Bills of Material Management (Arena Solutions, 2011).*

> *Since the late 1990s, outsourcing has become a way of life for electron-
> ics manufacturers. Most OEMs no longer consider manufacturing to
> be a core competency. Even in cases where some of this capability is
> retained in-house, there is an ongoing effort to evaluate more activi-
> ties that can be offloaded to a contract manufacturer (CM). These
> CMs, whose role in the electronics industry was previously limited to
> assembling printed circuit boards, have transformed themselves into
> large scale manufacturing powerhouses. Modern CMs provide their
> customers with a one-stop shop solution, providing excellence not only
> in manufacturing, but also in materials management, design and test
> services, order fulfillment, and logistics.*

It is not easy to determine where a supplier ends and a contract manufacturer
begins, so I have included the following definition from Wikipedia to further
illuminate the subject.

> *In a contract manufacturing business model, the hiring firm approaches
> the contract manufacturer with a design or formula. The contract
> manufacturer will quote the parts based on processes, labor, tooling,
> and material costs. Typically, a hiring firm will request quotes from*

multiple CMs. After the bidding process is complete, the hiring firm will select a source, and then, for the agreed-upon price, the CM acts as the hiring firm's factory, producing and shipping units of the design on behalf of the hiring firm.

The amount of oversight of the contract manufacturing company by the customer company can vary from little oversight to a great deal of oversight. No simple relationship can be assumed. Some contract manufacturers are larger than the companies that buy from them and tend to do more of the planning and maintain more control. In those situations, a passive form of planning would fit the business requirement. However, there are also cases where the customer company wants a high degree of control over the slave plant, to the point of actually performing production scheduling, which can be performed quite effectively by SAP APO.

If a customer company wants to perform detailed scheduling of the slave plant, they would implement PP/DS or another production planning and scheduling tool, and essentially provide the run sequence to the slave plant. The slave plant can be given access to the PP/DS planning board, check it, and simply make the schedule determined by the customer company. However, this scenario is rare.

The more common scenario is for the customer company to perform either active or passive supply planning and initial production planning, and to allow the slave plant to perform its own production scheduling. Under this scenario, purchase requisitions are created between the internal location and the vendor location (in APO), and these purchase requisitions initiate planned orders at the vendor location. The easiest way for people with some SNP exposure to understand this relationship is to think of it as if the requisition (which in this case would be a stock transport requisition) were being sent between two internal locations. Everything is the same, including where the requisition appears (in the Distribution Receipt key figure in the planning book at the internal location, and the Distribution Demand key figure in the planning location at the vendor location).

Is there any value in going into the vendor planning book product locations if they are planned with a finite method like CTM or the optimizer? There can be value. While the requisitions or planned orders would not necessarily be altered in the vendor product locations, the planning book can still serve as a valuable reporting tool. For instance, one could select any aggregation of vendors and quickly see the planned orders and purchase requisitions per period. Creating a selection profile in the planning book for all product and vendor location combinations would enable this type of reporting. Of course, any subset of product or vendor location could also be viewed in a similar manner.

In APO there are two basic ways to model vendors/suppliers, one of which is the standard way. I discovered the second way by working on projects and it is not well-explained.

1. Modeling the Supplier Location as an Internal Location

2. Emulating Supplier Capacity with GATP

I explain these two methods of modeling vendors/suppliers in the following paragraphs.

The Standard Approach: Modeling the Vendor Supplier as an Internal Location

SAP does not allow R/3 to place resources into a supplier. Secondly, SAP does not allow PDSs to be placed in supplier locations in APO. However, the resources in supplier locations can be modeled if they are set up as internal locations in SNP, and if these internal locations use PPMs. (For example, instead of being set up as a location type 1001 Vendor, they would be set up as a location type 1001 Production Plant. See the following article on the different location types in APO:

http://www.scmfocus.com/sapplanning/2012/07/05/location-types-in-sap-apo/)[10]

What this means is that the PPMs (resources, routings, and BOMs) must be maintained in APO rather than as production versions in R/3 and CIFed over, as is normally the case. In fact, the resources, routings, and BOMs cannot exist

---

[10] This brings up other complexities as well. Instead of purchase requisitions being generated, APO will generate stock transport requisitions as the vendors are modeled as internal locations or distribution centers. However, they can be created as billing STRs.

in SAP ERP because SAP ERP only sees vendor locations as sources of supply, which is to contain this as production information, and not as a location.

Focus On   The Evolution of the PDS

In covering this topic, I am describing the current situation with regards to APO in version 7.1. I have heard from consultants who work for SAP that, in the future, PDSs will be able to be adjusted from within APO—something that cannot be done currently and is a significant disadvantage to PDSs. This is particularly true when using a prototype to determine where it might be beneficial to be able to quickly create master data to build models. If eventually PDSs can be created and changed in APO, then it will be possible to use PDSs to model suppliers rather than PPMs. It will not change the fact that work centers/resources and production versions cannot exist in supplier locations in SAP ERP, but will simply allow PDSs to be used instead of PPMs.

For companies that have decided to use PPMs rather than PDSs, the following article explains when to use each:

http://www.scmfocus.com/sapplanning/2009/04/24/pds-vs-ppm-and-implications-for-bom-and-plm-management/

Companies that choose to use PPMs will experience less of a change than companies that choose to go with PDSs. Companies that use PPMs would designate each vendor location as simply another location that would use a PPM. Most likely, they would continue to manage true internal production locations with production versions in ERP, and then manage vendor location PPMs in APO. Companies that use PDSs will find this design to require more maintenance, because now they must maintain both PDSs for internal production locations and PPMs for vendor locations, meaning that those who use the system must become familiar with both objects.

For years SAP has been trying to get companies to move to PDSs, making the argument that the PDS provides better lifecycle capabilities (until it became clear that no one seems to use PDSs in this way). The current argument for PDSs is that they are supported for the long term, while PPMs are not. In general, SAP's arguments for using PDSs over PPMs have never been coherent, and a number of their early statements about the PDS turned out to not be true. However, SAP has settled on the argument that PDSs are supported more and allow companies to access more functionality than do PPMs. I have worked on a number of projects where the entire PDS versus PPM discussion has been a distraction. Until the PDS can be created and adjusted in APO rather than created in SAP ERP and brought over to APO, SAP has generated a permanent need for the PPM, even if it is not chosen as the main manufacturing master data object for the internal locations.

Common Modeling Approaches to Vendors/Suppliers

Production takes place in internal locations and in vendor/supplier locations. In most cases, product that is procured externally will be in vendors/suppliers that are not modeled as locations in SNP. When this is the case, the lead-time is taken from the Procurement tab of the Product Location Master.

In most cases, just a portion of the overall vendor database is modeled as locations. Doing so requires more effort to set up and manage, and is only beneficial for certain vendors, such as those with the largest volumes or with the most critical products. Secondly, there are several ways to model vendors, the most important distinction being whether the vendor is capacity-constrained. In cases where the vendor is not capacity-constrained (which could be because of the supply planning method used, or because of an active choice to not capacity-constrain the vendors), then the vendor location is used more for capacity visibility.

The options for capacity-constraining a vendor location are mostly the same as for an internal location, with fewer options for the technical objects that can represent the capacity.

http://www.scmfocus.com/sapplanning/2012/10/12/capacity-constraining-supplier-location-in-snp/

However, I have also come across large companies with few vendors, who did model every single vendor as a location. In general, this is feasible only for those companies with few vendors.

Technical Setup of the Solution in SNP

The technical setup for this solution starts at the resource. When internal locations are planned in APO, the BOM, routing and work center/resource is set up in ERP, and then brought into APO through the CIF to create either a PPM or a PDS (master data objects that combined the BOM, routing and work center/resource). ERP does not allow external locations to contain BOMs, routing and work centers/ resources. However, APO does allow this. Secondly, between the PPM and PDS, only the PPM can be created within APO. Therefore, when one needs to set up resources in an external location, the PPM can be created in APO but without representing the resource at the vendor location in ERP. Purchase requisitions are converted to purchase orders between the internal location and the vendor location, and capacity management at the vendor location becomes an entirely APO-related affair. Essentially, ERP never sees this setup, and in fact does not need to.

Routes between the vendor locations and the internal customer company locations can be created automatically as transportation lanes in APO. These transportation lanes are created from the purchasing information record in ERP. While it is not mandatory that all vendor locations be connected to internal locations with a transportation lane, this option does have some advantages with respect to consistency. However, transportation lanes do require more maintenance than the second option, which is not modeling a location, but using a planned delivery time on the Product Location Master. APO can use either the transportation lane or the Planned Delivery Time field, which is on the Procurement tab of the Product Location Master to calculate the transportation lead-time.

Supplier/Vendor Locations

Vendors can be modeled, or not, as locations. Vendors are set up as locations when the company wants to do the following:

1. View supplier capacities.

2. Constrain the vendor production on the basis of constrained resources.

3. Use the transportation load builder to build loads from the vendor to the company's own internal locations.

Disadvantages of the "Standard Approach"

The problem with actually modeling supplier resources in a location, which is this standard approach, is that it creates a lot of overhead. Companies have enough problems keeping their own resources up to date. Secondly, the effort required to set up this design greatly limits the number of suppliers that can be modeled. However, it is not the only option for modeling supplier capacity.

Emulating Supplier Capacity with GATP

Supplier capacity can be emulated with GATP through the use of allocations. Allocations stop a sales order from being accepted or from being confirmed if it is over the allocation. Allocations are much easier to set up than all of the master data objects that must be configured for SNP and PP/DS, thus allowing more suppliers to be "modeled" in the planning application and making the overall supplier modeling process much more sustainable.

This issue of sustainability is quite important because most companies that I have seen attempt supplier capacity modeling with the "standard approach," fail to pull it off successfully. SAP APO does make the standard approach possible, and SAP will often write it up as a way to design the supplier modeling solution, leaving out how difficult the process is to accomplish. They also do nothing to make the master data objects in APO easier to maintain. In fact, SAP plans to extend the ability to model supplier resources to the PDS, increasing the alternatives for companies that implement SNP.

GATP has the following interesting characteristics:

1. CIF synchronizes the data between SAP ERP and GATP in near real time.

2. The ATP check is performed in ERP when GATP is not available.

3. GATP integrates closely with supply planning for the order fulfillment process and commits to a customer's orders based on the input information provided by the supply planning receipt elements.

4. Four workflows connect supply planning or SNP to GATP:
   a. (GATP Check) ATP quantity (developed from inventory or MPS)
   b. (GATP Check) Sourcing (based upon product/location substitution: –GATP checks procurement for itself in a way that is not part of the supply plan)
   c. (GATP Check) Allocations management
   d. (GATP Check) Supply plan (daily/weekly MRP run)

However, the allocations we are thinking of creating are not connected to SNP. If we placed the resources and PPMs/PDSs into SNP, we would rely upon SNP to tell GATP what plan can be available. This design has GATP working off of allocations that are entered into the system and that serve as a proxy for the resources of the suppliers. This gets into the next topic, which is setting up allocations in GATP and how they are assigned to the product-location combination.

http://www.scmfocus.com/sapplanning/2012/08/21/using-gatp-allocations-for-modeling-supplier-capacity/

Allocation creation in GATP is quite flexible. Using it to model supplier as well as subcontractor capacity requires less maintenance than when the capacity is modeled in SNP. Performing constraint-based planning with anything but a small number of suppliers requires a significant amount of overhead, making GATP an important alternative. In fact, even unconstrained, it is quite a bit of work to manage supplier constraints through the standard process of setting up suppliers as internal locations, modeling their production lines, and then updating the production line capacity. More details on the GATP design to model supplier capacity are provided in the article below.

http://www.scmfocus.com/sapplanning/2012/08/21/using-gatp-allocations-for-modeling-supplier-capacity/

While the SNP-based design essentially takes the hard way, the GATP approach to supplier capacity constraints is the easier way. Instead of applying updated capacity information to the resource in the PPM, it is applied directly to the allocation in GATP. Of course the allocations must also be produced. This can be accomplished by setting target stock levels at the internal locations, which receive stock purchase orders to the supplier that match the allocation quantities. These target stock levels would be set in a way consistent with the desired allocations. The allocations functionality in GATP is very flexible; allocations can be applied to a combination of a customer and product, for just a product, for a product group, etc. The allocation, in effect, becomes a proxy for the production constraint. In fact, GATP has so many ways of setting up allocations and being configured, that the biggest challenge with GATP is simply making a decision of how to set it up.

The GATP approach is both feasible and lower maintenance when compared to the traditional SNP-centric approach to "modeling" supplier constraints. However, for whatever reason, this approach is not recommended frequently to clients. It may be that constraint-based planning is the province of SNP and PP/DS. It may also be that the allocation approach differentiates the management of the internally planned production locations differently from supplier production locations. However, this should not be a concern. It had occurred to me that one could manage all locations—both internal production locations and external production locations—in this same way. But, I am not familiar with a reference account for this design.

Technical Collaboration Challenges with SAP

In this chapter, I have spent a good deal of time describing the pros and cons of various modeling decisions that must be made when modeling supplier resources; however, these are not the only technical challenges. While it is often listed under business challenges, one of the issues affecting the propensity of companies to collaborate is whether or not the technology enables collaboration. Books on SAP do not address this topic, nor have I read any books on collaboration that have addressed it. In my view, this is because it is a universal requirement that authors talk up the technology rather than provide a more objective explanation. However, the truth is that SAP generally, as well as SAP APO, is rated poorly for collaboration. In my book *The Bill of Materials in Excel, ERP, Planning and PLM/BMMS Software*, I write in great length about a company called Arena Solutions, which sets the gold standard in terms of collaboration. The difference between Arena Solutions and what SAP offers is a yawning chasm. SAP does have collaborative products, such as the widely discussed but very lightly implemented SAP SNC (Supplier Network Collaboration), but none of their products really enable collaboration. SAP continues to promote collaboration, but many of its collaboration products such as Netweaver and HANA simply don't exist in any meaningful way.

http://www.scmfocus.com/sapprojectmanagement/2010/07/netweaver-does-not-exist/

http://www.scmfocus.com/scmbusinessintelligence/2011/11/is-gartner-now-distributing-sap-press-releases-as-analysis/

SAP's problems with collaboration are multi-dimensional and are described in the article below:

http://www.scmfocus.com/supplychaincollaboration/2010/06/flaws-in-saps-collaboration-technology/

There is a second problem with obtaining resource information from suppliers. Due to SAP's limited security model, it's difficult to provide suppliers with access to update the resource values directly in the system. Arena Solutions can do this easily—not with resources, as they don't make planning products, but with their specialty, the bill of materials.

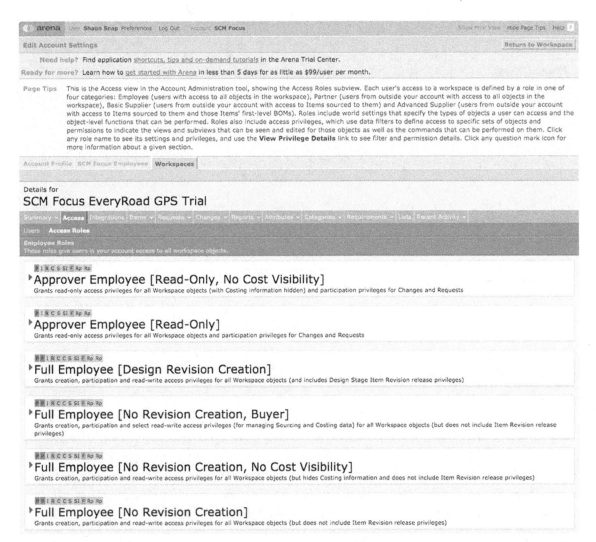

The colored boxes above the role (or the black and white boxes, if you are reading the printed edition of this book) define many different types of access. Because a fine level of control is allowed, access rights can be distributed broadly, both inside and outside the company that runs the application. This type of security, along with having a well-designed and easy-to-use user interface and with being natively web-based (SNC has several HTML "punch out" screens, but most of the application is built on the internally-oriented SAPGUI), means that the application is capable of truly supporting collaboration.

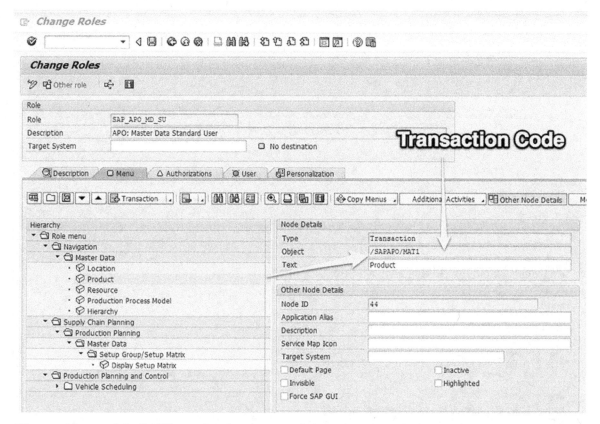

The problem with SAP's authorization model is that it is based upon the transaction. Authorization objects are below the transaction, but they take more effort to configure and to maintain. Because new users are being added perpetually and adjustments must be made in the security setting for other factors, a good security design must allow for changes to be made easily. This is covered in the following article.

http://www.scmfocus.com/sapplanning/2012/12/13/security-authorizations-in-sap-apo/

The Business Challenges of Modeling Supplier Capacity

Projects that model supplier capacity share similar business challenges to all other forms of supply chain collaboration (and perhaps non supply chain collaboration as well, but as I don't work outside of supply chain planning I can't say for sure). Many factors reduce the effectiveness of collaboration.

The book *Supply Chain Brutalization,* by Walt Grishcuk is quite illuminating about many aspects of contract manufacturing, including how badly contract manufacturers are often treated by original equipment manufacturers (OEMs). First it's important to consider how little CMs are paid. For instance, while Foxconn employs around half as many people as Walmart, their revenue is only fifteen percent of Walmart's revenue (as the Chinese currency is undervalued by roughly one half, on the basis of purchasing power parity, we might adjust this value up to thirty percent). Foxconn profit margin is roughly 3.7 percent, which is quite common in the contract manufacturing industry. There are many reasons as to why the profit margins are so low for CMs. Some of the profit is simply stripped out by OEMs, both through the contract negotiations and also by forcing the CMs to hold inventory without compensation when things are slow. As pointed out by Walt Grishuck, poor planning by the OEM can be pushed on to the CM. In fact, many OEMs use "lean" terminology in order to force suppliers to do things that are bad for the supplier but good for the OEM, as is brought up in the book *Supply Chain Brutalization.*

> *There are several ways to achieve cost reductions. Earlier we mentioned the brutal ways in which SCI was tormented each quarter by Racal. Is it safe to say that they would not have achieved the price points if they weren't so tenacious? Yes. But, look at the total costs, including the loss of revenues, which also occurred due to poor execution of the team and the resultant lack of desire of the CM to service them (business at a loss). Then add the costs associated with moving their manufacturing to a new supplier.*

> *We now operate in a brutal environment where the OEMs look for the company that can produce the products that will pass inspection for the lowest cost. This results in the sourcing of manufacturing in places that have the lowest labor rates, sometimes less than humane treatment to its work force, and the least amount of regulations on how the enterprise operates in its community or affects its environment.*

It's very difficult to see how "collaboration" can occur in such an environment. However, mistreating suppliers and then asking them to enter information in a collaboration system happens all too often. Poor treatment of suppliers not only reduces collaboration, but also reduces the incentive for sophisticated supply chain planning. It's interesting that so few books discuss this topic, but Walt's is one that does, and does so in a very honest fashion. No doubt it won't sell as well as books that promote "twenty-first century-based collaboration," because most readers prefer fantasy over reality and a simple storyline that only discusses the upside. This is the approach of many consulting companies as well. They simply sell the potential of collaboration. That is the best approach to achieving maximum sales.

Conclusion

Capacity-constraining suppliers is a very popular topic in companies, and is a growing requirement as companies outsource more of their manufacturing, and manufacturing is seen decreasingly as a core area of the business. However, there are very significant barriers to modeling supplier capacity, not the least of which is that outsourced manufacturers are different organizations and are often quite geographically remote from the company procuring the manufactured items. While the latest article from the *Harvard Business Review* or white paper from Booz Allen Hamilton may be quite positive about the potential for modeling supplier capacity, the results are uniformly worse for supplier capacity modeling than when one company manages the process internally.

Capacity-constraining suppliers can be both passive and active. Passive constraining is the most basic and tends to be used when the supplier is an "ordinary" supplier. There are two basic ways in APO to model vendors/suppliers. One is the standard way, and the second is a way that is not generally explained but that I discovered while working on projects.

1. Modeling the Supplier Location as an Internal Location

2. Emulating Supplier Capacity with GATP

When modeling suppliers in APO, there are some important distinctions to be aware of regarding whether to use PPMs or PDSs to model the supplier, and how supplier locations are modeled in APO versus SAP ERP. As noted above, the second approach to modeling supplier capacity in APO is with the use of allocations in GATP. Allocations stop a sales order from being accepted, or from being confirmed if it is over the allocation. Allocations are much easier to set up than all of the master data objects that must be configured for SNP and PP/DS when using the first option above—the "standard approach." This issue of sustainability is quite important because most companies that I have seen attempting supplier capacity modeling with the standard approach fail to pull it off successfully.

The security and authorizations model in SAP make it less likely that collaboration will occur, and therefore less likely that the resource information can be brought into SAP without an unreasonable amount of effort. When companies go down the path of modeling supplier capacity, they should be aware of these inherent collaborative limitations. These limitations are related to SAP's fundamental design and need to approve the appropriate amount of resources to overcome these issues. A company called Arena Solutions sets the standards in terms of security, and truly enables collaboration by making it easy to manage and change security functionality.

CHAPTER 7

The Disconnection Points Between Supply Planning and Production Planning

The previous chapter explained how APO follows a sequential approach, where the supply planning application and the production planning application hold different assumptions regarding time orientation, level of aggregation, and the existence of control factors such as labor pools and changeover matrixes which exist in PP/DS but not in SNP. In this chapter we will delve into these areas of disconnection between supply and production planning.

Changeovers in PP/DS

PP/DS has something called a Setup Matrix. The Setup Matrix holds the product-to-product downtimes for all the products that can be produced on a resource. A Setup Matrix looks like this:

New Setup Transitions

Location   SNA
Setup Matrix   SNA   Setup Matrix for SNA

Setup Transitions

	Predecssr	Successor	Setup Time	U..	Setup Costs
	Milk Product 1	Milk Product 5	25	H	400
	Milk Product 5	Milk Product 1	45	H	500

I quote from the book *Production Planning in SAP APO* on the topic of the Setup Matrix.

> *The setup matrix is defined in the master data of production planning and cannot be transferred from ECC (SAP ERP). The setup matrix is a transition matrix that contains the setup length of every possible setup transition. The setup length is needed to take the resource from one setup state to another. It can also contain setup costs that might be relevant to optimization. To use sequence dependent setup times, you must enter the setup matrix in the independent resource. You must also enter the setup key in the PPM/PDS for the relevant operations or enter the setup group, which you can transfer from the ECC routing. In addition, you must also set the setup activity flag in the setup activity so that the setup time is read from the setup matrix.*

A cost optimizer, like the one in APO, uses costs to help determine the manufacturing sequence, while the downtime is incorporated into the production plan. However, with a duration optimizer such as the one used in PlanetTogether, the duration is minimized as part of the objective function. Although Production Planning in SAP APO states that, "You can optimize the production plan in light of the total setup times or the total setup costs," I have not seen this done by clients. The book makes the further point:

Minimizing the total of the setup costs is necessary only when the system needs to have the costs and setup times described independently of each other. This can be the case when identical setup times differ on various resources in terms of cost-related evaluations and must therefore be decoupled from the setup time. Otherwise, the setup time criterion is sufficient.

I am going to have to disagree with this statement. But first, let's take a closer look at what this paragraph says; there is some important content if you read between the lines. It essentially states that costs can serve as a "proxy" for time and that the PP/DS optimizer can simply optimize on the costs within the Setup Matrix, meaning that it will attempt to minimize changeovers to create an optimal production schedule. The author would say that this is "coupling" the setup time with the setup costs. However, the costs will not be the actual costs of performing the changeover, but instead will be costs developed with very little logic supporting them (at least if the projects I have seen combined with numerous data points from discussions on this topic with other consultants is representative), and will certainly not be proportional to the other costs used by the PP/DS optimizer. But here is the problem, which I think is unaddressed in the quotation above: What should these costs be? To develop the right costs, a number of simulations will have to be run and the results shown to planners until PP/DS provides the output that the planners approve. I find this approach to controlling an optimizer to be very unscientific and a time waster, which is one reason I am not a big fan of cost optimizers in general (I have had good experiences with many optimizers, but not cost optimizers).

In my view, duration optimizers are superior to cost optimizers for production planning because durations are much easier for companies to determine than costs. This is because durations can be easily measured and have fewer shades of grey than actual costs. Durations have far fewer shades of grey than the made-up costs that drive the cost optimizer for both SNP and PP/DS; at least I have found this to be the case on every project I have been associated with. As the following article discusses, using cost optimization for every supply chain planning domain is a feature specific to first-generation supply chain planning optimizers that came out in the 1990s.

http://www.scmfocus.com/supplyplanning/2011/07/10/customizing-the-optimization-per-supply-chain-domain/

The PP/DS Setup Matrix can, in theory, be used to direct PP/DS to schedule production in a way that both meets demand and minimizes the manufacturing downtime while switching between different products. A Setup Matrix includes the durations (and costs when a cost optimizer is used as described in the article below) for changing the manufacturing line production from product to product. Each product-to-product changeover also has a cost associated with it, and because the PP/DS cost optimizer's objective function is to minimize costs, it seeks to schedule the planned orders in a way that minimizes the costs of the changeovers. This is just a broad-brush description, and more detail on this topic is available in the following article.

http://www.scmfocus.com/productionplanningandscheduling/2010/12/06/changeover-planning-in-sap-ppds-vs-planettogether/

The book *Production Planning in SAP APO* makes the following observation about which planning method to use if changeovers are a very important part of the planning process for a particular manufacturing environment.

> *If the setup problem of the bottleneck resource is not that important, or if it is of minor importance, scheduling the operations and orders is usually a straightforward process, because you can use the scheduling sequence in the planning strategy....if setup matrices are used and the total setup time accruing in a production program is largely dependent on the scheduling sequence due to the setup-condition dependent changeovers, in most cases the PP/DS Optimizer is the most effective tool for planning.*

So, according to the above quote, the optimizer is the best choice as a method if changeovers are a significant factor for production planning and scheduling.

> *Also, one or several alternatives should be available for bottleneck and the planning should be in a quantity and period oriented way*

(in other words, not order oriented). The latter ensures that planned orders can be split as required, should this be necessary due to the resource situation. This is often the case in the area of repetitive manufacturing. In such an environment, use of the PP/DS Optimizer is often not required. Instead, the multi-resource planning can be performed using a special heuristics, the wave algorithm.

This type of information is extremely helpful to companies that are choosing a planning method. In fact, two of the most important factors to determining the success of a project is both matching the software to the requirement, as well as matching the method to the requirement. Unfortunately, neither the software nor the methods are matched to the requirement anywhere near as often as they easily could be. Rather than performing the type of nuanced analysis above, the PP/DS optimizer may be selected for more superficial reasons.

Because the changeover matrix is used in PP/DS but generally not SNP, the results from SNP will simply be different from PP/DS. In essence, SNP sees both a higher capacity to resources than really exists, and believes that there is no penalty for any manufacturing sequence that it creates. How frequently the setup matrix is used by SNP is a topic which is difficult to get a firm answer.

Using the Setup Matrix in SAP SNP?

At every account I have seen the setup matrix used in PP/DS, but never by SNP. This is a major issue, although very lightly discussed on projects, because if PP/DS uses an assumption that SNP does not use, then the initial production plan that is sent from SNP to PP/DS would assume more capacity than PP/DS has available to it. Since SNP looks out further than PP/DS (SNP tends to have a planning horizon of a year, while PP/DS tends to have a horizon of 2 to 4 weeks), this would mean that the supply plan would bring in material with inaccuracies. However, while SNP has had this capability since version 5.1, I don't see clients use the setup matrix with SNP. SAP states that the following steps are necessary in order to use the setup matrix in SNP. I have shortened them a bit.

1. Create Setup Groups. To do so, on the SAP Easy Access screen, choose Advanced Planning and Optimization Master Data Application-Specific

Master Data Production Planning Setup Group/Setup Matrix Maintain Setup Groups.

2. The production data structure (PDS) used must contain a setup group.

3. Setup groups in SNP are supported by PDS only.

4. Define the setup statuses: set up transitions and the corresponding setup costs and setup times in the setup matrix. To do this, on the *SAP Easy Access* screen, choose *Advanced Planning and Optimization Master Data Application-Specific Master Data Production Planning Setup Group/Setup Matrix Maintain Setup Matrix*.

5. Specify the setup matrix in the master data of a single-mixed resource or a bucket resource on the *Time-Cont. Capacity* tab page. Other resource types are not supported.

6. Set the *Per Lot Size* indicator for the resource to *Sequence-Dependent Lot-Size Planning* on the *SNP Bucket Cap* tab page.

7. A PDS can use only one resource for which this indicator has been set.

8. A PDS must not exceed any planning period (bucket) within the defined time period for sequence-dependent lot size planning.

9. In the SNP optimizer profile, select *Discrete Optimization* as the optimization method and, in the *Sequence-Dependent Lot Size* field on the *Discrete Constraints* tab page, specify a time period in which the SNP optimizer takes sequence-dependent lot sizes into account.

However, I am a bit suspicious of this functionality as it is so little discussed on projects.

Production Planning in SNP with Cycle Planning/Periodic Lot Sizing

Cycle planning is a way for SNP to make adjustments that indirectly incorporate changeovers into the SNP-generated (i.e., initial) production plan. That is, these adjustments help to account partially for the reduced number of projected available production hours associated with changeovers, as well as for other factors that reduce the resource's capacity.

The first step to this adjustment is to use a manufacturing cycle. A manufacturing cycle is the frequency with which production is allowed and is most often set on a weekly time basis. Manufacturing cycles can be set up in SNP, per product, or per product group (by using the grouped product functionality in SNP) within the Lot Size tab of the Product Location Master. This cycle plan sets the allowable weeks that a product could be produced. So if a product could be produced in Week Three, but not Week One or Two, the product would be on a three-week cycle. In order for the cycle plan to be effective, analysis must be performed to assign complementary products into the same cycle. Thus the optimizer can choose to schedule production for complementary products in the same week, but may not do the opposite. That partially helps account for the Setup Matrix in PP/DS, but it does not account for the reduced resource availability time for production when the resource is down during a changeover.

1. *A Period Type: Defines the type of period (for example, day, week, month, and so on) to be used for period lot-sizing.*

2. *The Number of Periods: Defines the number of periods that are to be grouped together when forming lot sizes. You only need to define this field for period lot-sizing procedures.*

3. *The Planning Calendar (for Lot Sizing Procedure): For the periodic lot-sizing procedure, use an APO standard time stream as a planning calendar. This field is not taken into account in Supply Network Planning (SNP).*

4. *The Start Time for Grouping Horizon: This indicator is only relevant for heuristic-based planning in Supply Network Planning (SNP). It is used together with the period lot-sizing procedure, which you set using the Period Lot Size indicator. You can specify when you want the system to start grouping the demands together (i.e., when the grouping horizon starts). The following options are available:*
   a. *Blank: The grouping of demands starts in the first period of the planning bucket profile defined in the planning book (as a rule, today's date).*
   b. *X: The grouping of demands starts in the first period of the planning bucket profile where there is a supply shortage.*

This is another example of where a planning calendar/time stream is assigned to an object—in this case, the Lot Size tab of the product location master.

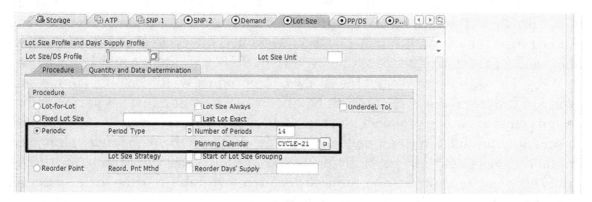

The periodic lot sizing manufacturing is timed off of a time stream/planning calendar.

When the periodic lot size/manufacturing cycle is set up, it can be configured to show up in a key figure in the planning book, which is named "No Production." If a "1" exists in the key figure, then the system cannot schedule production, and if a "0" exists in the key figure, then the system can schedule. However, it should be emphasized that a manual planned order can be entered in any planning bucket regardless of whether it is coded with a zero or not.

Adjusting Supply Planning for Labor Pool Functionality in Production Planning

Labor pools were explained in Chapter 2: "Understanding the Basics of Constraints in Supply and Production Planning Software." As discussed in that chapter, labor pool functionality exists in PP/DS and SAP PP, but does not exist in SAP SNP, meaning that SNP creates its initial supply plan (within which is the initial production plan) without incorporating any labor pools—which may be in PP/DS or PP (both have labor pool functionality). Therefore, PP/DS and PP have information that SNP cannot see and which changes the resulting plan. PP/DS will change the sequence and timing, as well as the quantities of the planned orders that were generated by SNP (when they fall into the PP/DS Planning Horizon). Depending upon the length of the lead times, the company will procure material for a production schedule that is bound to change, and depending upon circumstances, could change significantly. The end result is a planning process that is much less accurate than it could be if the supply planning application worked off of the same assumptions as the production plan.

Conclusion

There are a number of points of disconnect between supply planning and production planning applications in general and between SNP and PP/DS specifically. These disconnects include the following:

1. PP/DS has a changeover/Setup Matrix, which reduces the useful working time of resources, but SNP does not.

2. PP/DS has labor pools that serve as a co-capacity constraint with resources, and SNP does not.

As a result, SNP and PP/DS will calculate different output because they can work on different assumptions. I say "can" work on different assumptions because PP/DS can be implemented without a Setup Matrix and without a labor pool. However, there are some changes that can be put into effect in SNP to close the gap between SNP and PP/DS. Cycle planning is a way of emulating the PP/DS Setup Matrix in SNP by combining complementary products together in weeks so that changeovers are minimized.

CHAPTER 8

Conclusion

The objective of constraint-based planning is to generate a feasible supply and production plan. Constraint-based planning restricts activities to declared limits as specified in resources. Resources can be anything from factory equipment to transportation units to handling equipment, but for both supply planning and production planning in most cases they are production resources. The system then moves loads, locating and relocating capacity automatically during the planning run. Constrained-based planning is a relatively recent development for off-the-shelf supply chain planning software that is commercially available. It has been roughly 20 years since this type of software was first implemented in industry, but few companies have mastered constraint-based planning, even those that have live and constrained versions of SNP, PP/DS or other supply planning and production planning applications.

The opposite of constraint-based planning (or finite planning) is capacity leveling or infinite planning. Capacity leveling or infinite planning allows any load to be placed onto any period. Capacity leveling is then applied in a second step to smooth out the peaks and valleys. Not every resource type can actually be capacity-leveled in APO. Capacity leveling only works in SNP for production resources and transportation resources, *and does not work for storage or handling resources*. Furthermore, capacity leveling only works on four of the resource types: bucket, single-mixed, multi-mixed, and transportation.

This book described heuristics, CTM, and optimization, which are the different methods that can be used for supply and production planning in APO. Using either CTM or optimization software allows for the inclusion of constraints that are not restricted to the supply planning domain, such as constraints taken from production planning. Discrete optimization is a more challenging form of optimization that results in more realistic planning output, but also in less optimal, or higher cost, output. However, the point of any modeling exercise is to make the model as accurate as is reasonable (not as "possible," because increased model accuracy requires more costs, both in the short term and in the long term), not to attain an optimal solution based upon the most simplified assumptions. Methods that perform constraint-based planning are often described as "constraint-based methods," the implication being that they are exclusively used for constraint-based planning. However, the truth is much greyer than that. Constraint-based methods can perform constraint-based planning, but they are not always set up to do so. Secondly, any of the constraint-based methods can be used with some resources that are constrained and some that are not. This is an important distinction because it's quite common to not only have mixed constrained resources, but resources that are constrained to change depending upon whether the resource is planned by the live version versus the simulation version.

Resources are the mechanism for both constraining the plan in SAP APO and determining the feasibility of the plan. SNP can use all the resource types (production resources in addition to supply planning resources), while PP/DS can only use production resources. Placing resources into a supply planning system is the next step up from using simple lot sizes in the emulation of production constraints. Because of the nature of some manufacturing processes, they are often sequential. Constraint-based applications for production planning leverage this fact into modeling only the bottleneck resource. However, supply planning resources—and supply planning in general—is less sequential in nature, and some of the resources (such as transportation resources or handling resources) are not constrained in the same way as production resources. Resources tend to be the focus of constraint-based planning and capacity leveling; however, there are several other important constraints including the Setup Matrix and the resource calendars. Some companies use a co-constraint called a labor pool, which is functionality in PP/DS and is a way of restricting the scheduling of production based upon machine capacity and on the available labor in the factory.

The hand-off between SNP and PP/DS is quite important, and the topic of planning horizons in both SNP and PP/DS were covered for this reason. In terms of timing, the trick is to configure the various horizons so that SNP and PP/DS are responsible for creating planned (production) orders for different times—although as was explained in this book there can be an overlap in this time between SNP and PP/DS.

A number of companies that I have worked with have requested capacity constraining of suppliers. One can capacity-constrain internal production locations, but to determine overall feasibility it is necessary to capacity-constrain suppliers as well, particularly the largest suppliers. Because of changes to the business environment (where manufacturing is increasingly outsourced), the requirement to either constrain or have visibility into suppliers has greatly increased in the past fifteen years. SAP does not allow R/3 to place resources into a supplier. Secondly SAP does not allow PDSs to be placed in supplier locations in APO. However, the resources in supplier locations can be modeled if they are set up as internal locations in SNP, and if these internal locations use PPMs. Production takes place in internal locations and in vendor/supplier locations. A second approach to modeling suppliers is by emulating them with GATP through the use of allocations. Allocations stop a sales order from being accepted, or from being confirmed if it is over the allocation. Allocations are much easier to set up than all of the master data objects necessary to configure for SNP and PP/DS.

References

Babu, Mahesh. *SCM7.0 PPDS: Cross Plant Deployment (New Functionality)*. SAP, 2010.

Balla, Jochen and Layer, Frank. *Production Planning in SAP APO* (2nd Edition), SAP Press, 2010.

Beyond BOM 101: Next Generation Bills of Material Management, Arena Solutions, 2011. http://www.arenasolutions.com/resources/articles/bill-of-materials.
http://www.sdn.sap.com/irj/scn/go/portal/prtroot/docs/library/uuid/20049b9f-e542-2d10-a083-d3cf60dde64e?QuickLink=index&overridelayout=true&47515223697288.

Bottleneck. Last modified February 25, 2013. http://en.wikipedia.org/wiki/Bottleneck.

Capacity Leveling. http://help.sap.com/saphelp_ewm70/helpdata/en/90/d9733b570d474bba36bf443f7927c0/content.htm.

Chiu, Alex. *Multi-Plant Scheduling*. Last modified November 18, 2011. http://www.help.apsportal.com/advanced-topics/multi-plant-and-multi-user-settings/multi-plant-scheduling.

Chiu, Alex. *Plant Stability*. Last modified November 18, 2011. http://www.help.apsportal.com/advanced-topics/multi-plant-and-multi-user-settings/plant-stability.

Comparison of the Capacity Leveling Methods.
http://help.sap.com/saphelp_ewm70/helpdata/en/e3/e9cc13badd4344a3e0aefb
6b9ac265/content.htm.

Contract Manufacturer. Last modified February 26, 2013.
http:// en.wikipedia.org/wiki/Contract_Manufacturer.

Dickersbach, Jorg Thomas, Gerhard Keller and Klaus Weihrauch. *Production Planning and Control with SAP ERP*, SAP Press, 2010.

Gaddam, Balaji. *Capable to Match (CTM) with SAP APO.* SAP Press, 2009.

Grischuk, Walt. *Supply Chain Brutalization: The Handbook for Contract Manufacturing.* BookSurge Publishing, 2010.

Hill, Arthur V. *Encyclopedia of Operations Management.: A Field Manual and Glossary of Operations Management Terms and Concepts.*, FT Press, 2011.

Interior Point Method. Last modified March 7, 2013. http://en.wikipedia.org/wiki/Interior_point_method.

Klier, Thomas and James Rubenstein. *Who Really Made Your Car? Restructuring and Geographic Change in the Auto Industry.* W.E. Upjohn Institute, 2008.

Michels, Patrick and Edson Bacin. *Enabling Lean Supply Chain Planning in SAP APO.* Paper presented at the ASUG Annual Conference, Orlando, FL, May 15-18, 2011.
http://events.asug.com/2011AC/1205_Enabling_Lean_Supply_Chain_Planning_
in_SAP_APO.pdf.

Opportunity Cost. Last modified February 28, 2013.
http://en.wikipedia.org/wiki/Opportunity_cost.

Optimization-Based Planning.
http://help.sap.com/saphelp_ewm70/helpdata/en/09/707b37db6bcd66e10000009b
38f889/content.htm.

Planning with Aggregated Resources.
http://help.sap.com/saphelp_ewm70/helpdata/en/43/03b0b3dccd22f3e10000000a
1553f7/frameset.htm.

Planning with Bottleneck Resources.
http://help.sap.com/saphelp_scm40/helpdata/en/e8/c6765e14e84890b8ce7c0
cf7f29384/content.htm.

Pradham, Sandeep and Pavan Verma. *Global Available-to-Promise with SAP: Functionality and Configuration.* SAP Press, 2011.

Resource Disaggregation.
http://help.sap.com/saphelp_ewm70/helpdata/en/43/03b26bdccd22f3e10000000a
1553f7/content.htm.

Setup Matrix
http://help.sap.com/saphelp_ewm70/helpdata/en/47/fca2fd9fc4384be10000000a
42189b/content.htm.

Simplex Algorithm. Last modified February 25, 2013. http://en.wikipedia.org/wiki/
Simplex_algorithm.

Snapp, Shaun. *The Bill of Materials in Excel, ERP, Planning and PLM/BMMS
Software.* SCM Focus Press, 2013.

Stadtler, Hartmut and Bernhard Fleishmann. *Advanced Planning in Supply Chains.*
Springer Press, 2012.

Stock Transfers with PP/DS. In SAP Community Network forum, January 15, 2009.
http://scn.sap.com/thread/1195661.

Vendor Acknowledgements and Profiles

I have listed brief profiles of each vendor with screen shots included in this book below.

Profiles:

PlanetTogether

PlanetTogether's software is not for supply planning but instead for production planning and scheduling. However, they have been included in this book to demonstrate principles of supply chain optimization. PlanetTogether, software developer of Galaxy APS (Advanced Planning & Scheduling), enables manufacturers to eliminate spreadsheets and connect multi-plant operations for real-time visibility and collaboration. Users report fifty percent reductions in inventory and labor costs, faster time-to-delivery, and a six-month return on investment. PlanetTogether serves nearly one hundred clients in process and discrete manufacturing in a broad range of vertical segments.

www.planettogether.com

SAP

SAP does not need much of an introduction. They are the largest vendor of enterprise software applications for supply chain management.

SAP has multiple products that are showcased in this book, including SAP ERP and SAP APO.

www.sap.com

Author Profile

Shaun Snapp is the Founder and Editor of SCM Focus. SCM Focus is one of the largest independent supply chain software analysis and educational sites on the Internet.

After working at several of the largest consulting companies and at i2 Technologies, he became an independent consultant and later started SCM Focus. He maintains a strong interest in comparative software design, and works both in SAP APO, as well as with a variety of best-of-breed supply chain planning vendors. His ongoing relationships with these vendors keep him on the cutting edge of emerging technology.

Primary Sources of Information and Writing Topics

Shaun writes about topics with which he has first-hand experience. These topics range from recovering problematic implementations, to system configuration, to socializing complex software and supply chain concepts in the areas of demand planning, supply planning and production planning.

More broadly, he writes on topics supportive of these applications, which include master data parameter management, integration, analytics, simulation and bill of material management systems. He covers management aspects of enterprise software ranging from software policy to handling consulting partners on SAP projects.

Shaun writes from an implementer's perspective and as a result he focuses on how software is actually used in practice rather than its hypothetical or "pure release note capabilities." Unlike many authors in enterprise software who keep their distance from discussing the realities of software implementation, he writes both on the problems as well as the successes of his software use. This gives him a distinctive voice in the field.

Secondary Sources of Information

In addition to project experience, Shaun's interest in academic literature is a secondary source of information for his books and articles. Intrigued with the historical perspective of supply chain software, much of his writing is influenced by his readings and research into how different categories of supply chain software developed, evolved, and finally became broadly used over time.

Covering the Latest Software Developments

Shaun is focused on supply chain software selections and implementation improvement through writing and consulting, bringing companies some of the newest technologies and methods. Some of the software developments that Shaun showcases at SCM Focus and in books at SCM Focus Press have yet to reach widespread adoption.

Education

Shaun has an undergraduate degree in business from the University of Hawaii, a Masters of Science in Maritime Management from the Maine Maritime Academy and a Masters of Science in Business Logistics from Penn State University. He has taught both logistics and SAP software.

Software Certifications

Shaun has been trained and/or certified in products from i2 Technologies, Servigistics, ToolsGroup and SAP (SD, DP, SNP, SPP, EWM).

Contact

Shaun can be contacted at: shaunsnapp@scmfocus.com or www.scmfocus.com/

Abbreviations

APO (SAP) – Advanced Planning and Optimizer

APS – Advanced Planning and Scheduling

ATD – Available to Deploy

ATP – Available to Promise

BOM – Bill of Materials

CIF – Core Interface

CTM (SAP) – Capable to Match

DC – Distribution Center

DP (SAP) – Demand Planner

DRP – Distribution Resource Planning

ERP – Enterprise Resource Planning

GATP (SAP) – Global Available to Promise

MASSD – Not an acronym, but the mass maintenance transaction in SAP APO.

MDM – Master Data Management

MEIO – Inventory Optimization and Multi-echelon Planning

MOT – Means of Transport

MPS – Master Production Schedule

MRP – Materials Requirements Planning

OEM – Original Equipment Manufacturer

PDS – Production Data Structure

PP/DS (SAP) – Production Planning and Detailed Scheduling

PPM – Production Process Model

RDC – Regional Distribution Center

SAPGUI – SAP Graphical User Interface

SNC (SAP) – Supplier Network Collaboration

SNP (SAP) – Supply Network Planning

S&OP – Sales and Operations Planning

SD (SAP) – Sales and Distribution

STO – Stock Transport Order

STR – Stock Transport Requisition

TOC – Theory of Constraints

TP/VS – Transportation Planning and Vehicle Scheduling

VMI – Vendor Managed Inventory

Links Listed in the Book by Chapter

Chapter 1

http://www.scmfocus.com/writing-rules/

http://www.scmfocus.com

http://www.scmfocus.com/supplyplanning

http://www.scmfocus.com/productionplanningandscheduling/

http://www.scmfocus.com/sapplanning

Chapter 2

http://www.scmfocus.com/supplyplanning/2011/07/10/customizing-
the-optimization-per-supply-chain-domain/

http://www.scmfocus.com/sapplanning/2011/10/12/snp-optimizer-
sub-problem-division-and-decomposition/

http://www.scmfocus.com/inventoryoptimizationmultiechelon/
2011/05/socializing-supply-chain-optimization/

Chapter 4

http://www.scmfocus.com/sapplanning/2009/05/06/pegging-in-scm/

http://www.scmfocus.com/sapplanning/2008/09/21/ppds-and-snp-heuristics/

http://www.scmfocus.com/sapplanning/2008/09/21/ppds-and-snp-heuristics/

http://www.scmfocus.com/sapplanning/2011/06/09/forecast-consumption-setting/

http://www.scmfocus.com/sapplanning/2012/06/27/backward-scheduling-forward-scheduling-sap-erp-sap-apo/

http://www.scmfocus.com/sapplanning/2012/06/13/front-loading-resources-in-sap-snp/

http://www.scmfocus.com/inventoryoptimizationmultiechelon/2011/10/redeployment/

http://www.scmfocus.com/sapplanning/2012/07/04/product-location-database-segmentation-for-snp-and-supply-planning/

http://www.scmfocus.com/sapplanning/2008/09/21/ppds-and-snp-heuristics/

http://www.scmfocus.com/sapplanning/category/ppm-pds/

http://www.scmfocus.com/sapplanning/2013/02/19/discrete-optimization-horizon/

http://www.scmfocus.com/sapplanning/2012/07/27/the-connection-between-boms-routings-work-centers-in-erp-and-ppms-pdss-in-apo/

http://www.scmfocus.com/sapplanning/2012/08/08/labor-pool-as-a-constraint-in-snp-and-ppds/

http://www.scmfocus.com/sapplanning/2012/06/13/front-loading-resources-in-sap-snp/

http://www.scmfocus.com/sapplanning/2013/01/31/version-copy-in-apo/

http://www.scmfocus.com/sapplanning/2009/05/06/pegging-in-scm/

http://www.scmfocus.com/sapplanning/2012/06/22/order-or-supply-and-demand-selection-for-ctm/

http://www.scmfocus.com/sapplanning/category/ctm/

http://www.scmfocus.com/sapplanning/2011/06/09/forecast-consumption-setting/

Chapter 5

http://www.scmfocus.com/sapplanning/2009/05/02/scm-resource-types/

http://www.scmfocus.com/sapplanning/2012/12/06/the-goods-receipt-processing-time-and-the-handling-resource/

http://www.scmfocus.com/sapplanning/2012/08/19/time-continuous-planning-versus-bucket-in-ctm-and-ppds/

http://www.scmfocus.com/sapplanning/2012/06/05/ctm-customer-priorities-versus-order-priorities/

http://www.scmfocus.com/sapplanning/2009/12/08/customer-prioritization-and-ctm/

http://www.scmfocus.com/productionplanningandscheduling/2010/12/06/changeover-planning-in-sap-ppds-vs-planettogether/).

http://www.scmfocus.com/supplychaininnovation/2009/06/google-maps-and-gomobileiq-for-vehicle-routing/

http://www.scmfocus.com/fourthpartylogistics/2012/04/the-overestimation-of-outsourced-logistics/

http://www.scmfocus.com/productionplanningandscheduling/2012/07/04/who-was-the-first-to-engage-in-mass-production-ford-or-the-venetians/

http://www.scmfocus.com/scmhistory/

http://www.scmfocus.com/sapplanning/2011/11/05/how-soft-constraints-work-with-soft-constraints-days-supply-and-safety-stock-penalty-costs/

Chapter 6

http://www.scmfocus.com/sapplanning/2012/07/05/location-types-in-sap-apo/

http://www.scmfocus.com/sapplanning/2009/04/24/pds-vs-ppm-and-implications-for-bom-and-plm-management/

http://www.scmfocus.com/sapplanning/2012/10/12/capacity-constraining-supplier-location-in-snp/

http://www.scmfocus.com/sapplanning/2012/08/21/using-gatp-allocations-for-modeling-supplier-capacity/

http://www.scmfocus.com/sapplanning/2012/08/21/using-gatp-allocations-for-modeling-supplier-capacity/

http://www.scmfocus.com/sapprojectmanagement/2010/07/netweaver-does-not-exist/

http://www.scmfocus.com/scmbusinessintelligence/2011/11/is-gartner-now-distributing-sap-press-releases-as-analysis/

http://www.scmfocus.com/supplychaincollaboration/2010/06/flaws-in-saps-collaboration-technology/

http://www.scmfocus.com/sapplanning/2012/12/13/security-authorizations-in-sap-apo/

Chapter 7

http://www.scmfocus.com/supplyplanning/2011/07/10/customizing-the-optimization-per-supply-chain-domain/

http://www.scmfocus.com/productionplanningandscheduling/2010/12/06/changeover-planning-in-sap-ppds-vs-planettogether/